Library of Congress Cataloging in Publication Data

Aviado, Domingo M.
Non-fluorinated propellants and solvents for aerosols.

(Uniscience toxicology series: Solvents in the environment series)
Includes bibliographies and indexes.
1. Solvents – Toxicology. 2. Trichloroethane – Toxicology. 3. Dichloromethane – Toxicology. 4. Alcohol – Toxicology. 5. Aerosol propellants – Toxicology. 6. Propane – Toxicology. 7. Butane – Toxicology. 8. Methylpropane – Toxicology. I. Zakhari, Samir, joint author. II. Watanabe, Tetsuya, 1947– joint author. III. Title.
RA1270.S6A94 615.9'51 77-736
ISBN 0-8493-5199-5

International Standard Book Number 0-8493-5199-5

Library of Congress Card Number 77-736
Printed in the United States

Non-fluorinated Propellants and Sol for Aerosols

Authors

Domingo M. Aviado
University of Pennsylvania
School of Medicine
Philadelphia

Samir Zakhari
University of Pennsylvania
School of Medicine
Philadelphia

Tetsı
Unive
Sc

Editor-in-Chief

Leon Golberg

Chemical Industry Institute of Toxicology
Research Triangle Park, North Carolina

Solvents in the Environment Series
in
Uniscience Toxicology Series

Published by

CRC PRESS, Inc.
18901 Cranwood Parkway · Cleveland, Ohio 44128

OTHER TITLES IN THE CRC SOLVENTS IN THE ENVIRONMENT SERIES

Editor-in-Chief, **Leon Golberg, M.S., B.Chir., D.Sc., D.Phil., F.R.C.Path.**

METHYL CHLOROFORM AND TRICHLOROETHYLENE IN THE ENVIRONMENT

By **Domingo M. Aviado, M.D., Samir Zakhari, Ph.D.,** both at the University of Pennsylvania School of Medicine; **Joseph A. Simaan, M.D.,** American University of Beirut; and **Andrew G. Ulsamer, Ph.D.,** United States Consumer Product Safety Commission.

ISOPROPANOL AND KETONES IN THE ENVIRONMENT

By **Zamir Zakhari, Ph.D., Mitchell Liebowitz, Ph.D., Paul Levy, B.A.,** and **Domingo Aviado, M.D.,** all at the University of Pennsylvania School of Medicine.

This monograph is dedicated to Dr. Henry Field Smyth, Jr.

FOREWORD

Henry Smyth, Jr. ("Smyth as in myth") is not everybody's cup of tea! In an age where the trend is towards smooth, oleaginous public relations types, he is the very antithesis – blunt, critical, abrasively intolerant of fools or knaves, the epitome of a man of science. His long and distinguished career spans years of work in toxicology and occupational health. For 30 of those years he was the Director ("Senior Fellow") of the Carnegie-Mellon Institute at Bushy Run, Pittsburgh. From his pen and those of his distinguished colleagues flowed a series of reports on investigations in virtually every aspect of industrial toxicology and many other areas of safety evaluation of a wide range of chemicals.

Throughout his life, Henry Smyth has enriched toxicology by his propensity for deep and original thinking – not an invariable attribute, even among the more senior toxicologists of our time. I still recall the impact of my first meeting with him, at a Gordon Conference in Toxicology, when he delivered his memorable address on the subject of "Sufficient Challenge" (*Food Cosmet. Toxicol.*, 5, 51, 1967). This masterly essay is well worth reading today, especially by the johnny-come-latelies who think that the purpose of toxicology should be to insure absolute safety, or at least to get rid of all those "chemicals" that defile the pristine character of the "natural" environment.

Dr. Henry Field Smyth, Jr. was born in Germantown, Pennsylvania, on June 5, 1903. His father started as a country doctor but later headed the Department of Hygiene in the School of Medicine at the University of Pennsylvania. Father and son became partners in a consulting practice which provided industrial hygiene and toxicological services for industry on the Eastern Seaboard, with numerous publications reflecting the broad scope of their pioneering studies. Dr. Smyth, Jr. likewise joined the faculty of the Department of Hygiene where he taught Bacteriology and Food, Air and Water Chemistry.

In 1967 he retired from the Chemical Hygiene Fellowship of Mellon Institute to reinaugurate a teaching career at the Graduate School of Public Health, University of Pittsburgh. Dr. Smyth has received the Donald E. Cummings Award, American Industrial Hygiene Association 1956; the Society of Toxicology Merit Award 1966; and the Mellon Award in 1976 – all evenly spaced in his characteristically orderly fashion. During his entire career Dr. Smyth has devoted a monumental amount of effort to further the American Industrial Hygiene Association, as Executive Secretary, President, and member of the American Board of Industrial Hygiene.

It is with respect and admiration that the authors and I dedicate this volume to Henry Smyth, Jr. Men of his ability and intellectual stature are becoming increasingly scarce in the field of toxicology. Never have they been more sorely needed than now!

Leon Golberg
Research Triangle Park, North Carolina

PREFACE

As this third volume, *Non-fluorinated Propellants and Solvents for Aerosols,* was being written, a first volume on *Methyl Chloroform and Trichloroethylene* had been published and another on *Methylene Chloride* was in preparation. The present volume does not deal with a single chemical but with three solvents and three propellants that are likely to be used in aerosol products. Although it is a common practice to combine solvents and propellants in pressurized containers, there has been no previous attempt to examine each constituent singly and in combination.

This volume contains the results on the investigation of the inhalational toxicity of the following chemicals: methyl chloroform (Chapter 1), methylene chloride (Chapter 2), ethanol (Chapter 3), propane (Chapter 4), butane (Chapter 5), and isobutane (Chapter 6). Each chapter is complete with a review of the literature with the exception of methyl chloroform and methylene chloride which are covered in other volumes. The interaction among the chlorinated solvents and hydrocarbon propellants (Chapters 7, 8, and 9) concludes this volume.

Domingo M. Aviado

THE EDITOR-IN-CHIEF

Leon Golberg is currently President of Chemical Industry Institute of Toxicology, Research Triangle Park, North Carolina. Dr. Golberg received his B.S., M.S., and D.Sc. from the University of Witwatersrand, Johannesburg, South Africa and his D. Phil. from the University of Oxford, England. Dr. Golberg has been a Consultant to the World Health Organization and has served the Food and Drug Administration, the National Institutes of Health, and the National Academy of Sciences in various capacities. He is currently Editor-in-Chief of the CRC Uniscience Toxicology Series and Editor of *CRC Critical Reviews in Toxicology.*

THE AUTHORS

Domingo M. Aviado is the Professor of Pharmacology and Toxicology at the University of Pennsylvania School of Medicine at Philadelphia. He obtained his M.D. degree from the same University in 1948 and immediately joined the staff of the Department of Pharmacology.

Samir Zakhari is an Instructor in Pharmacology at the University of Pennsylvania. He obtained his Ph.D. from the Slovak Academy of Science and is on leave of absence from the Pharmacology Unit, National Research Center of Cairo.

Tetsuya Watanabe received a D.D.S. degree in 1971 from Kanagawa Dental College of Japan. He is now an Associate in the Department of Pharmacology of the University of Pennsylvania School of Medicine.

ACKNOWLEDGMENTS

We would like to thank Dr. Herbert L. Ratcliffe and David L. Snyder for their assistance in the preparation of Chapters 1, 7, and 8. The original investigation described in Chapters 1 to 8 has been supported, at least in part, by funds from the United States Consumer Product Safety Commission (CPSC-C-75-078) and the Food and Drug Administration (FDA 223-71-2310). The content of this monograph does not necessarily reflect the views of the Commission or the Administration, nor does the mention of trade names, commercial products, or organizations imply their endorsement.

TABLE OF CONTENTS

Part I
Solvents for Aerosols

Chapter 1
METHYL CHLOROFORM

Among the solvents discussed in the earlier volumes of this series,[1,2] methyl chloroform has been selected as the prototype of a chlorinated solvent widely used in aerosol products. The review of the literature[1] revealed that methyl chloroform, when introduced as a general solvent, was a significant improvement over the more toxic chlorinated solvents such as carbon tetrachloride, chloroform, and trichloroethylene. The lower level of toxicity of methyl chloroform compared to other chlorinated solvents was confirmed in the original investigation[1] describing acute toxicity in mice and dogs.

The experiments that are reported below are essentially subacute in nature. Since fluorocarbons administered for brief periods (30 min) daily for 3 weeks produced bronchopulmonary functional abnormalities and histopathological lesions in mice,[3] it became necessary to examine the non-fluorinated solvents, specifically methyl chloroform. This was the initial step in the examination of aerosol solvents and propellants that are not fluorinated. In the event that fluorocarbon propellants are banned because of their suspected effect of depleting the ozone layer, substitutes that are non-fluorinated are under consideration and methyl chloroform may serve the purpose of a low-pressure propellant replacing trichloromonofluoromethane (FC 11). In the classification of aerosol propellants and solvents proposed in 1974, the two compounds were grouped together as interchangeable, and methyl chloroform is less toxic than FC 11 in provoking cardiac arrhythmia, depressing the myocardial contractility and reducing systemic arterial blood pressure.[4]

However, the potential hepatotoxicity of methyl chloroform appears to be more serious than that of FC 11. The experiments reported below include functional, histological, ultramicroscopic, and chemical analytic examination of the liver of mice using an exposure schedule that would simulate the incidental inhalation of aerosol. The protocol previously used for subacute fluorocarbon inhalation has been employed: 30 min inhalation daily for 1, 2, or 3 weeks.[3]

A. Methods

Male mice of the inbred strain Balb-C were purchased at weights of 20 to 25 g from the Charles River Breeding Laboratories (Wilmington, Massachusetts) and separated into 11 groups each comprising 14 animals. They were maintained as separate groups in laboratory cages with a pelleted commercial ration and water *ad libitum*.

1. Exposure Chamber and Exposures

Inhalation exposures were made in a 38.5-l glass chamber with a flow of 30 l/min. Liquid methyl chloroform was vaporized at 15°C, 8.5°C below room temperature, to produce a saturated vapor without condensation. Before being drawn into the chamber (temperature 23.5°C) the saturated gas was mixed with a column of room air, concentration being continuously monitored by an infrared gas analyzer. The room air-solvent mixture was drawn through the chamber under negative pressure and exhausted through an activated charcoal filter.

The mice were exposed to three concentrations of the solvent: 500, 1000, and 2000 ppm, during three intervals: 90 min/day, 6 days/week for 1 week; 60 min/day, 6 days/week for 2 weeks; and 30 min/day, 6 days/week for 3 weeks. The 42 mice to be exposed to a given concentration were color-coded and all placed in the chamber at one time, and after 30 min 14 were returned to their cage. Again, after 60 min, another 14 were returned to their cage, and the exposure continued for another 30 min, the gas flow, of course, being stopped at the end of each interval until the mice had been removed.

2. Bronchopulmonary Functions

When exposures had been completed, nine mice from control groups and each exposure group were taken for measurements of tidal volume, pulmonary resistance, and pulmonary compliance by the following method: The mouse was anesthetized with sodium pentobarbital (25 mg/kg, i.p.). The trachea cannulated and a plastic catheter inserted into the pleural cavity in order to measure pressure change by means of a Validyne pressure transducer (model MP45-1). The mouse was then placed in the body plethysmograph containing a mesh screen pneumotachograph for measurement of respiratory movement. Pressure differences corresponding to air flow across the screen were measured by a Statham pressure transducer (model PM9TC) and then integrated into tidal volume. Pulmonary compliance (tidal volume/transpulmonary pressure) was computed at zero flow between inspiration and expiration. Pulmonary resistance was calculated by the following equation: $P_1 - P_2/F_1 - F_2$. P_1 and P_2 are intrapleural pressures at isovolumetric points on either side of the respiration cycle and F_1 and F_2 are flows at isovolumetric points. The values were recorded by a pulmonary mechanics computer (Buxco Electronics). The details of the bronchopulmonary techniques have been described in an earlier publication from this laboratory.[5]

3. Liver: Total Lipids, Phospholipids, and Glycogen

When pulmonary function had been recorded each mouse was weighed, killed, and its liver quickly removed for a record of wet weight, then wrapped in aluminum foil and frozen at −15°C for later study. The whole liver (weights about 1 to 2 g) was homogenized in cold chloroform: methanol (2:1) the volume adjusted to 20 ml, transferred to a Corex® tube and centrifuged at 12,000 r.p.m. for 20 min. Ten ml of the supernatant was used for determining total lipids and phospholipids according to methods previously used in this laboratory.[5,6] The dried pellet was digested in 30% potassium hydroxide and its glycogen content measured by the method of Seifter et al.[7]

4. Tissues for Microscopic Study

The five mice remaining from each group were lightly anesthetized with ether, and the body cavities quickly opened. The liver was then removed and a part was taken for histological studies. Small segments were cut from the remainder, fixed for about 30 min in glutaraldehyde, then cut into 1- to 2-mm cubes and fixation continued for about 4 hr. The specimens were washed through two changes of 0.1 *M* cacodylate buffer and stored in a refrigerator. When processed for electron microscopy, they were dehydrated in ethanol-water solutions, embedded in Epon® 812 and cut with diamond knives at 75 to 90 nm.

Tissues for light microscopy were fixed in buffered neutral 10% formalin, embedded in paraffin, sectioned at 5 μm and stained with hematoxylin and eosin. Sections for fat stains were cut at 20 μm from frozen blocks and stained with oil red O and hematoxylin.

B. Results

The experiments were so designed that each of three concentrations of methyl chloroform (500, 1000 and 2000 ppm) might be used under the following three conditions of exposure:

90 min/day for 6 days (1 week)

60 min/day for 6 days/week for 12 exposures (2 weeks)

30 min/day for 6 days/week for 18 exposures (3 weeks).

Control groups were exposed to room air for 1, 2, or 3 weeks.

1. Bronchopulmonary Function

The measurements of pulmonary resistance, pulmonary compliance, tidal volume, and respiratory rate are summarized in Table 1.1. The daily exposures of 500, 1000, and 2000 ppm for up to 3 weeks did not influence any of the bronchopulmonary measurements. There was no difference from controls in the various parameters, indicating that exposure to methyl chloroform did not alter bronchopulmonary function.

2. Response to Hypoxia

The bronchopulmonary effects of a 30-sec exposure to 7.5% oxygen consist of increase in respiratory minute volume, decrease in pulmonary resistance, and increase in pulmonary compliance. This episode of hypoxia has been used previously to examine the influence of propellants on regulation of the airways.[3] The effects of hypoxia are summarized in Table 1.1.

After week 1 of exposure, none of the values were significantly changed, but after week 2, the percentage differences in pulmonary compliance were significantly reduced in mice on 1000 and 2000 ppm. At this time, too, the difference in respiratory rates was reduced in the 2000 ppm group. After week 3, all values except pulmonary resistances were within normal limits but in this instance, the percentage difference between controls and experimental mice, although significant, probably reflect an effect of anesthesia or other manipulations and may be considered to be an artifact.

Thus, pulmonary complicance, which may have reflected hyporeactivity, after week 1, as well as

TABLE 1.1

Functional Respiratory Changes in Mice Inhaling 7.5% Oxygen for 30 Sec After Exposure to One of Three Concentrations of Methyl Chloroform*

	Pulmonary resistance cm H_2O/ml/sec			Pulmonary compliance ml/cm H_2O			Tidal volume ml			Respiratory rate per min		
	Before	After	% Δ	Before	After	% Δ	Before	After	% Δ	Before	After	% Δ
Week 1												
Controls (6)	1.62 ± 0.069	1.46 ± 0.048	−9.6 ± 1.0	0.0242 ± 0.0009	0.0299 ± 0.0001	24.1 ± 3.73	0.134 ± 0.007	0.174 ± 0.0148	28.9 ± 4.10	116.7 ± 5.73	137.5 ± 7.72	17.7 ± 1.14
500 ppm (6)	1.68 ± 0.070	1.51 ± 0.058	−9.9 ± 0.73	0.0217 ± 0.0013	0.0261 ± 0.0013	20.8 ± 2.49	0.138 ± 0.0039	0.177 ± 0.0025	29.4 ± 3.62	105.8 ± 4.17	128 ± 6.41	21.1 ± 2.02
1000 ppm (6)	1.68 ± 0.057	1.51 ± 0.047	−10.1 ± 0.84	0.0219 ± 0.0008	0.0265 ± 0.0011	21 ± 2.34	0.143 ± 0.0061	0.186 ± 0.011	30.1 ± 4.06	119.2 ± 4.73	140.8 ± 5.07	18.3 ± 1.70
2000 ppm (7)	1.62 ± 0.048	1.46 ± 0.05	−9.7 ± 1.09	0.0237 ± 0.0009	0.0273 ± 0.0013	15.1 ± 2.60	0.131 ± 0.0072	0.166 ± 0.0135	25.1 ± 3.39	104.3 ± 3.85	123 ± 5.2	18.5 ± 2.4
Week 2												
Controls (9)	1.51 ± 0.05	1.45 ± 0.051	−9.4 ± 0.92	0.0240 ± 0.0009	0.0313 ± 0.00108	29.6 ± 1.5	0.159 ± 0.0044	0.220 ± 0.0093	38.4 ± 4.05	128.3 ± 5.0	155.0 ± 5.53	21.0 ± 1.03
500 ppm (7)	1.58 ± 0.03	1.41 ± 0.026	−10.9 ± 8.0	0.0248 ± 0.0009	0.0311 ± 0.00123	25.1 ± 2.12	0.139 ± 0.0059	0.191 ± 0.01	37.7 ± 2.03	114.3 ± 2.97	136.4 ± 4.46	19.4 ± 2.97
1000 ppm (7)	1.62 ± 0.04	1.46 ± 0.035	−9.8 ± 0.96	0.0242 ± 0.0017	0.0293 ± 0.00218	20.9 ± 2.26**	0.147 ± 0.0068	0.196 ± 0.015	32.8 ± 4.98	115.0 ± 1.88	136.4 ± 5.42	18.8 ± 1.65
2000 ppm (7)	1.53 ± 0.04	1.37 ± 0.04	−10.5 ± 1.17	0.0288 ± 0.00075	0.0348 ± 0.00097	21.1 ± 2.27**	0.172 ± 0.0062	0.23 ± 0.0075	33.4 ± 2.95	117.1 ± 4.06	136.3 ± 4.55	16.4 ± 1.52**
Week 3												
Controls (7)	1.69 ± 0.056	1.59 ± 0.053	−5.7 ± 0.55	0.0283 ± 0.0022	0.0345 ± 0.003	21.8 ± 2.97	0.161 ± 0.007	0.214 ± 0.011	32.9 ± 4.13	115.6 ± 3.95	131.9 ± 3.92	14.2 ± 1.49
500 ppm (7)	1.68 ± 0.058	1.50 ± 0.037	−10.5 ± 1.41**	0.0248 ± 0.009	0.0311 ± 0.0011	25.4 ± 2.08	0.148 ± 0.002	0.192 ± 0.004	29.4 ± 2.0	110.7 ± 4.14	132 ± 5.33	20.1 ± 2.58
1000 ppm (7)	1.66 ± 0.05	1.50 ± 0.05	−9.5 ± 1.0**	0.0220 ± 0.0006	0.0273 ± 0.0009	24.0 ± 2.89	0.138 ± 0.006	0.171 ± 0.007	24.9 ± 2.59	117.1 ± 3.91	137.1 ± 5.41	17.0 ± 1.57
2000 ppm (9)	1.71 ± 0.03	1.52 ± 0.03	−10.9 ± 0.28**	0.0231 ± 0.0007	0.0273 ± 0.0006	18.2 ± 2.17	0.131 ± 0.005	0.159 ± 0.006	21.9 ± 3.49	113.3 ± 3.23	129.4 ± 2.82	14.5 ± 1.79

*90 min/day, 6 days/week during week 1 60 min/day for 12 days weeks 1 and 2; 30 min/day, 18 days, weeks 1, 2, and 3.

**Indicates $P < 0.05$ difference between control and experimental values.

after week 2, showed a tendency to decreased values, although the difference from controls was not statistically significant except for week 2. After week 3 this tendency was not evident. This may be interpreted as evidence of adaptation.

3. Acute Response to 10% Methyl Chloroform

The results of a further test for changes in pulmonary function in the mice that inhaled methyl chloroform for 30 min/day, 6 days/week for 3 weeks are shown in Table 1.2. Here, values are given only as percentage differences in function before and after inhaling 10% of the chlorinated hydrocarbon for 2 min on the eighteenth day of exposure to the less concentrated gas. This table shows a decrease in pulmonary resistance in each exposure group which became significant only in mice on 2000 ppm, combined with a significant percentage difference in respiratory rate. This change is interpreted to mean decreased airway resistance, possibly caused by an anesthetic effect of the vapor.

4. Body Weights, Liver Weights, Lipids, Phospholipids, and Glycogen

Body weights, wet weights of livers, liver weight/body weight ratios, total liver lipids, total liver phospholipids, percent of phospholipid in total lipids, and values for liver glycogen for control mice and for each of the three groups that inhaled methyl chloroform are shown in Table 1.3 as group means after weeks 1 and 2 of exposure. Values after week 3 of exposure had returned to normal and are not included.

Values given in Table 1.3 show that after exposure-week 1, statistically significant changes ($p < 0.05$) in body weight, wet weight of liver, total phospholipids, and percent of phospholipids were limited to mice exposed to 2000 ppm of methyl chloroform. Glycogen also was increased in each of the exposure groups at this time.

After 2 weeks of exposure, each of these values, with the exception of liver glycogen for the 2000 ppm exposure group, corresponded closely to values for controls. Values for liver glycogen, however, were decreased compared to control values. These conflicting changes in values for liver glycogen cannot be explained at this time, but probably reflect an effect of food intake or lack of it.

5. Liver Histologic Changes Under Light Microscopy

Pathological changes that might be attributed to inhalation of methyl chloroform were limited to the liver. Table 1.4 lists the frequency of these changes as fractions of the number of mice exposed to one of the three concentrations for the stated intervals.

This table shows that all of the mice exposed to methyl chloroform had developed fatty livers (microglobular fat) during week 1, and the fatty livers persisted through weeks 1, 2, and 3 (Figures 1.1 and 1.2). However, fat in the livers of these mice cannot be attributed solely to inhalation of the hydrocarbon for the livers of more than half of the controls also contained fat in amounts that, by oil red O-staining, sometimes equalled that in the livers of the exposed mice. Fatty livers in the controls, at least, may be attributed to an in-

TABLE 1.2

Respiratory Functional Changes of Mice Exposed to 10% Methyl Chloroform for 2 Min (Mean ± SEM)*

Exposure (No. mice)	Pulmonary resistance $cm\ H_2O/ml/sec$ % Δ	Pulmonary compliance $ml/cm\ H_2O$ % Δ	Tidal volume ml % Δ	Respiratory rate % Δ
Control air (7)	−5.1 ± 0.20	+0.6 ±0.6	+0.7 ±0.7	+8.7 ±2.19
500 ppm (7)	−5.4 ± 0.25	+0.0 ±0	+0.0 ±0	+3.2** ±1.02
1000 ppm (7)	−6.8 ± 0.8	+0.0 ±0	+0.0 ±0	+2.6** ±0.94
2000 ppm (7)	−9.8 ± 0.7**	+0.0 ±0	+0.0 ±0	+1.1** ±0.58

*Superimposed upon 3 weeks' exposure to methyl chloroform, 30 min/day, for 6 days/week.
**Significant difference ($p < 0.05$).

TABLE 1.3

Hepatic Effects of Inhalation of Methyl Chloroform in Mice

Procedure (No. mice)	Body wt g	Wet liver wt g	Wet liver wt. / 100 g body wt.	Total lipids mg/g	Total phospholipids mg/g	% Phospholipids to lipids
	Mean ± SEM	Mean ± SEM	Mean ± SEM	Mean ± SEM	Mean ± SEM	Mean ± SEM
1 week exposure 90 min/day						
Control (9)	25.6 ± 0.78	1.27 ± 0.028	5.06 ± 0.096	56.5 ± 2.64	31.4 ± 0.45	56.19 ± 3.103
500 ppm (9)	24.3 ± 0.53	1.27 ± 0.032	5.24 ± 0.135	49.6 ± 2.37	30.95 ± 0.74	62.75 ± 2.79
1000 ppm (9)	24.4 ± 0.70	1.30 ± 0.061	5.31 ± 0.125	48.1 ± 2.64	30.28 ± 0.94	65.01 ± 3.072
2000 ppm (9)	22.1 ± 0.54	1.08 ± 0.024*	4.91 ± 0.104	60.2 ± 2.79	26.78 ± 0.34*	45.16 ± 2.21*
2 weeks exposure 60 min/day						
Control (9)	23.8 ± 0.52	1.31 ± 0.051	5.50 ± 0.164	54.7 ± 1.47	29.1 ± 0.59	54.45 ± 1.589
500 ppm (9)	22.9 ± 0.47	1.20 ± 0.028	5.21 ± 0.052	57.1 ± 1.05	29.9 ± 0.49	52.47 ± 1.003
1000 ppm (9)	23.4 ± 0.58	1.22 ± 0.062	5.21 ± 0.150	59.1 ± 2.47	29.3 ± 0.92	49.91 ± 2.145
2000 ppm (9)	23.9 ± 0.35	1.30 ± 0.029	5.43 ± 0.075	56.5 ± 1.77	29.3 ± 0.63	52.11 ± 1.584

*Significant difference ($p < 0.05$)

TABLE 1.4

Hepatic Changes Associated with Intermittent Inhalation Exposures to Methyl Chloroform*

Procedure		Hepatocytes: Cytoplasm: Fatty change			Hepatocytes: Nuclear changes			Biliary tract lesions
		Total	Centrolobular	Diffuse	Pyknosis	Necrosis	Mitoses	
Controls		6/10	0/10	6/10	0/10	0/10	0/10	0/10
500 ppm	90 min/day; 1 week	5/5	3/5	2/5	0/5	3/5	2/5	0/5
	60 min/day; 2 weeks	5/5	0/5	5/5	0/5	0/5	0/5	0/5
	30 min/day; 3 weeks	5/5	4/5	1/5	0/5	2/5	3/5	0/5
1000 ppm	90 min/day; 1 week	5/5	2/5	3/5	0/5	1/5	0/5	0/5
	60 min/day; 2 weeks	5/5	4/5	1/5	0/5	1/5	3/5	0/5
	30 min/day; 3 weeks	5/5	5/5	0/5	0/5	1/5	2/5	0/5
2000 ppm	90 min/day; 1 week	5/5	0/5	5/5	0/5	2/5	0/5	0/5
	60 min/day; 2 weeks	5/5	4/5	1/5	0/5	0/5	2/5	0/5
	30 min/day; 3 weeks	5/5	5/5	0/5	0/5	0/5	3/5	0/5

*Frequency of changes shown as fractions of numbers of mice exposed.

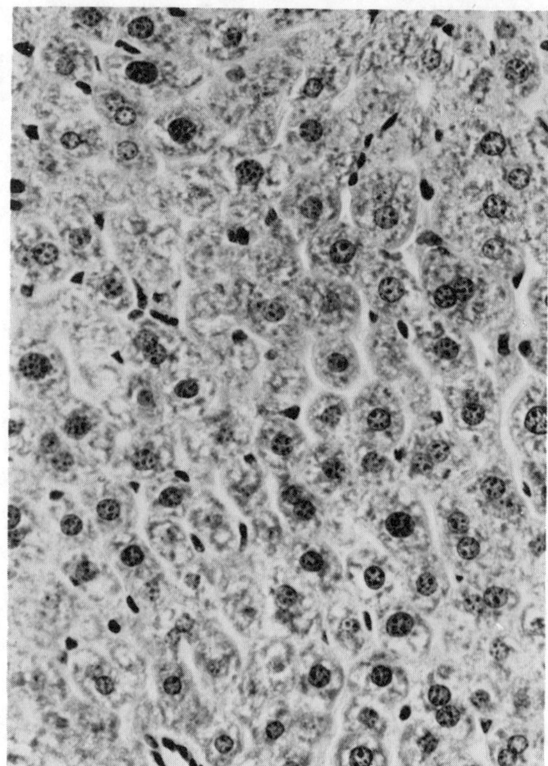

FIGURE 1.1. Periportal region of a liver lobule (portal zone at base of figure; note bile duct). Central vein out of field at top. Many cells contain small vacuoles. Control mouse tissue stained with hematoxylin and eosin (× 400).

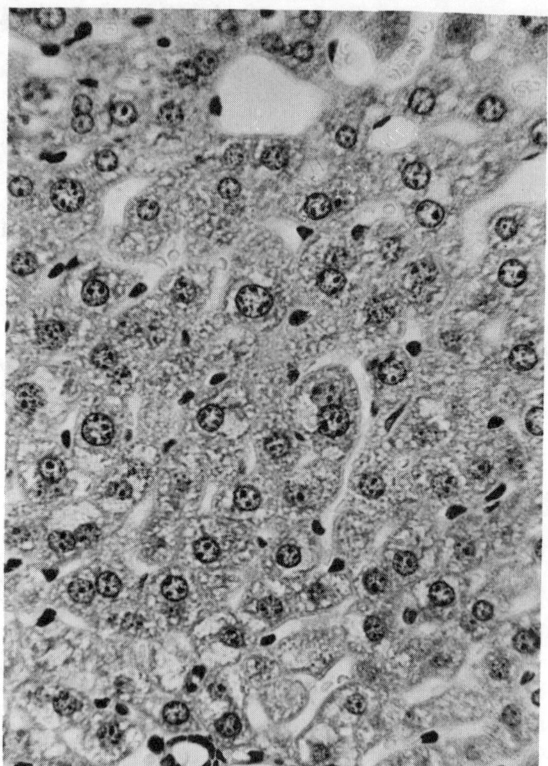

FIGURE 1.2. Part of a liver lobule, portal zone at base of figure; note bile duct, branch of a central vein at top. Many cells throughout the lobule, from portal to central zone, are vacuolated. Mouse after 1-week exposure to 2000 ppm methyl chloroform; tissue stained with hematoxylin and eosin (× 400).

completely balanced ration. Moreover, the fat droplets in the livers of controls were always diffuse, which was also the case with some, if not all, mice of the three exposure groups during week 1, but by week 3, the fat droplets had become largely centrolobular in distribution. Oil red O-stained preparations of livers from mice killed after week 3 suggested that the amounts of fat had been reduced in a majority, as evidence of recovery.

Mitoses were rare in the livers of control mice, being found in only one animal in which frequency was approximately 1 mitotic figure per 2000 nuclei. After week 1, mitoses were seen only in mice of 500 ppm of the vapor, two of five animals having 1 mitotic figure per 1000 nuclei. In mice on 500 ppm of the vapor relative numbers of mitoses were not increased, but in mice on 1000 ppm, they were increased in 3 of 5 mice after week 2 and in mice on both 1000 ppm and 2000 ppm after week 3.

Pyknotic nuclei were not recognized in any of the preparations, although in livers in which mitosis was seen, occasional nuclei stained very darkly. However, these nuclei were always larger than average and have been interpreted as an early stage in mitosis. Degenerating cells, or cells undergoing necrosis, were most numerous after week 1 and were found in the livers of mice exposed to each of the three levels of vapor, especially 500 ppm and 2000 ppm (Figures 1.3 and 1.4). These necrotic cells seemed to be randomly distributed, occurring singly, or in groups of two or three, as enlarged, pale-staining, relatively coarsely vacuolated bodies with pale-staining nuclei. Only a few examples (Figure 1.3) of eosinophilic cytoplasmic inclusions were found in these cells and they were inconspicuous. Mitoses of liver cells were seen only in exposed animals, and at the higher concentrations, only after weeks 2 and 3, which suggests recovery. Inflammatory responses to the degenerating cells, like hypertrophy and hyperplasia of

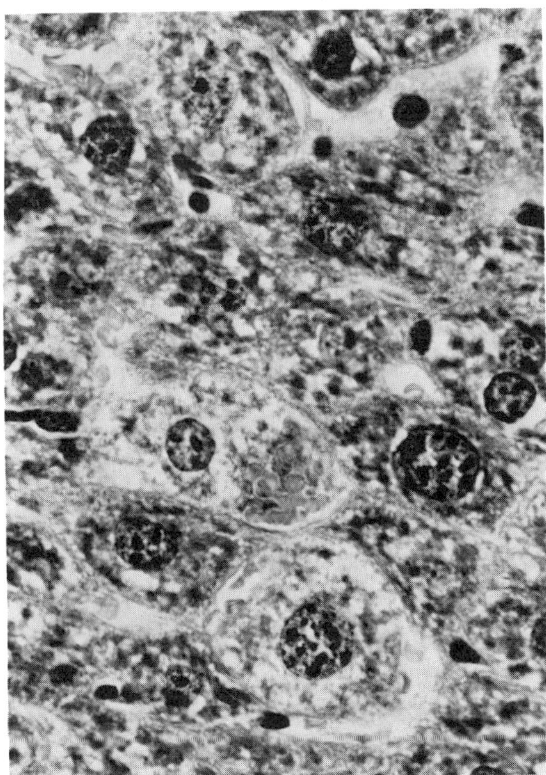

FIGURE 1.3. Two hepatocytes in lower half of figure are swollen and coarsely vacuolated; the upper cell of these two contains spherical eosinophilic masses. At middle top of the field a third hepatocyte also is undergoing degeneration. Its nucleus stains faintly, whereas nuclei of the other two cells stain more intensely. Mouse after 1-week exposure to 500 ppm methyl chloroform; tissue stained with hematoxylin and eosin (× 880).

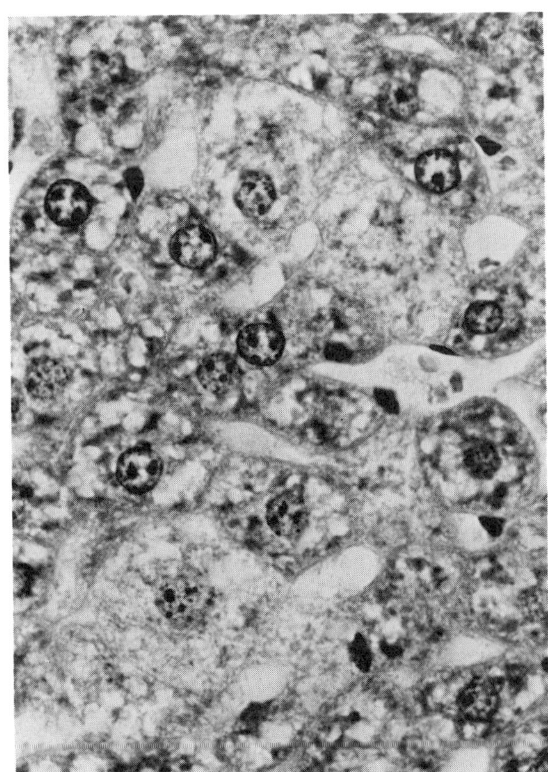

FIGURE 1.4. Hepatocyte necrosis, lower left and upper right of field; cells are enlarged, pale-staining, and coarsely vacuolated, their nuclei faintly staining. Surrounding cells are all vacuolated. Mouse after 3-weeks exposure to 2000 ppm methyl chloroform; tissue stained with hematoxylin and eosin (× 880).

Kupffer cells and inflammatory reactions in the biliary tracts were not recognized.

6. Electron Microscopy of Liver

Sections of liver from one control mouse at week 1, and from exposed mice, 500 ppm week 1 and 2000 ppm week 2, were examined by the electron microscope (Figures 1.5, 1.6, and 1.7). In so far as might be determined in these three animals by this method, there was no appreciable disturbance to the ultrastructure of the liver cells by inhalation exposure to methyl chloroform under the conditions of this experiment.

7. Histologic Changes in Other Visceral Organs

Sections of each lobe of the lungs were examined for changes in alveoli or in bronchial epithelium. Evidence of bronchial epithelial hyperplasia such as had been found in the lungs of Balb-C mice that had inhaled fluorocarbons,[3] were not observed here. Similarly, hearts, spleens, kidneys, and testes revealed no change that might be attributed to inhalation of methyl chloroform.

C. Discussion

The bronchopulmonary effects of the fluorocarbons have been recently elucidated by this laboratory.[3,4] Brief, 30-min daily exposures for 3 weeks caused pathological changes in the lungs, specifically hyperplasia of the lymphoreticular cells. There was also a disturbance in regulation of the bronchomotor tone, reflected by hyposensitivity to hypoxia. The experiments reported above indicate that there is no hyperplasia and no alteration in function of the airways in mice exposed to methyl chloroform.

The addition of liver studies to our experiments

was decided upon because, contrary to previously published data, Horiguchi and Horiuchi reported in 1971 that hepatic lesions resulted from 1-hr exposure on alternate days for a total of nine exposures to 1000 ppm 1,1,1,-trichloroethane.[8] There were no control animals in the experiments by Horiguchi and Horiuchi, and these authors did not illustrate in their publication the histologic changes described as "... congestion of the lungs and liver and traces of inflammation around the biliary duct area."

In the present study, there was no congestion of the lungs and liver or traces of inflammation around the biliary duct area. Necrosis of hepatocytes occurred in mice at each dilution of the compound at the end of week 1, but was not regularly distributed among the mice after week 2 or week 3 of exposure. Moreover, the extent of hepatocyte necrosis in the present material was equally pronounced at each dilution, and approximately equal in extent in all involved livers to that illustrated by McNutt et al.[9] However, hepatocyte necrosis in the present material was not accompanied by inflammatory responses nor by appreciable increases in Kupffer cells as seen by McNutt et al. following continuous exposure to 1000 ppm for 14 weeks.[9]

One explanation for these reports of inflammatory responses to hepatocyte necrosis, like the report of inflammation about the biliary tracts of mice exposed to this chlorinated hydrocarbon by Horiguchi and Horiuchi,[8] may be that the mice used in those studies carried chronic hepatitis which was activated by the exposures to methyl chloroform. Chronic hepatitis is a relatively common disease of laboratory mice.

McNutt et al.[9] have reported much more severe liver damage in mice, as seen by electron microscopy, after continuous exposure to 1000 ppm of methyl chloroform. We cannot explain the failure of mice in the present experiment to exhibit comparable liver changes unless it is a result of intermittent exposures.

The literature on the hepatotoxicity of methyl chloroform, summarized in Table 1.5, indicates that brief exposures do not influence the liver in the following species: 18,000 ppm in rats[10] and 2650 ppm in humans.[11] For daily exposure of up to 8 hr, the following concentrations were tolerated without influencing the liver: 5000 ppm in rats,[10] 2000 ppm in guinea pigs,[12] 5000 ppm in rabbits,[10] 2200 ppm in dogs,[13] and 5000 ppm in monkeys.[10] For continuous exposure, 24 hr daily, a level of 250 ppm for 14 weeks was tolerated by mice and rats, and 1000 ppm for 14 weeks by dogs.[12,14]

The concentration that consistently produces hepatic lesions is at least 1000 ppm continuous exposure. The only exception is the report of Tsapko and Rappoport stating that 73 ppm of 4 hr daily exposure produces swelling of liver cells and venous hyperemia in rats.[15] Since other investigators have failed to note these lesions in rats exposed to concentrations higher than 73 ppm, the unusual lesions reported by Tsapko and Rappoport are regarded as unique for their experimental conditions and strain of rat.[15]

D. Summary

Mice of the inbred strain Balb-C were caused to inhale methyl chloroform in concentrations of 500 ppm, 1000 ppm, and 2000 ppm for 90 min/day, 6 days/week, for 1 week, 60 min/day, 6 days/week for 2 weeks, and 30 min/day, 6 days/week for 3 weeks. Exposed mice, with controls, were examined for pulmonary function, lipid, phospholipid, and glycogen content of livers and for histological changes attributable to the exposures. Pulmonary function was unchanged. When tested with hypoxia, the response of the airways of the exposed mice was no different from the controls. When tested against 10% of the methyl chloroform for 2 min, mice of all groups after 3 weeks of exposure had decreased pulmonary resistance. Total liver lipids were increased in mice on 2000 ppm after 1 week, but not thereafter. Changes in glycogen content of livers were inconsistent. Hepatocyte necrosis was observed in all exposure groups after the first week, in the 1000 ppm group after each week, and in the 500 ppm group after the third week. Necrosis was random, affecting 2 to 3 cells in a given area, and not accompanied by inflammatory change or hyperplasia of Kupffer cells, nor were lesions of the biliary tracts seen. Liver changes, as demonstrated by electron microscopy, were minimal or absent.

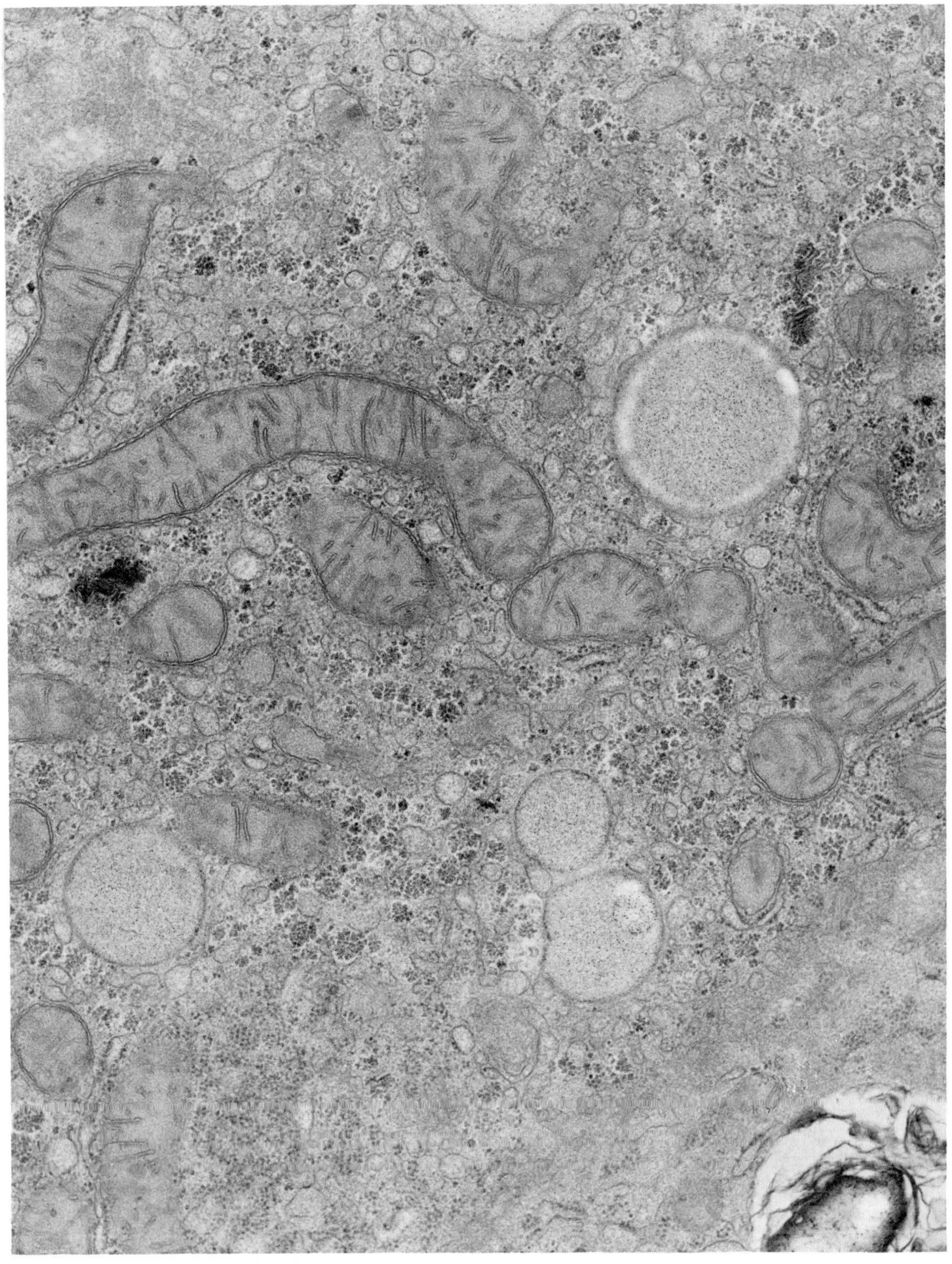

FIGURE 1.5. Electron micrograph of a portion of an hepatocyte from control mouse exposed to room air, 90 min/day, 6 days/week. Mitochondria prominent; few strands of rough endoplasmic reticulum. Near center right and lower left, four fat droplets stain lightly with OsO_4; smooth endoplasmic reticulum seen as circles and loops in several parts of field; glycogen abundant. Dark strands at lower and upper right not identified (× 31,000).

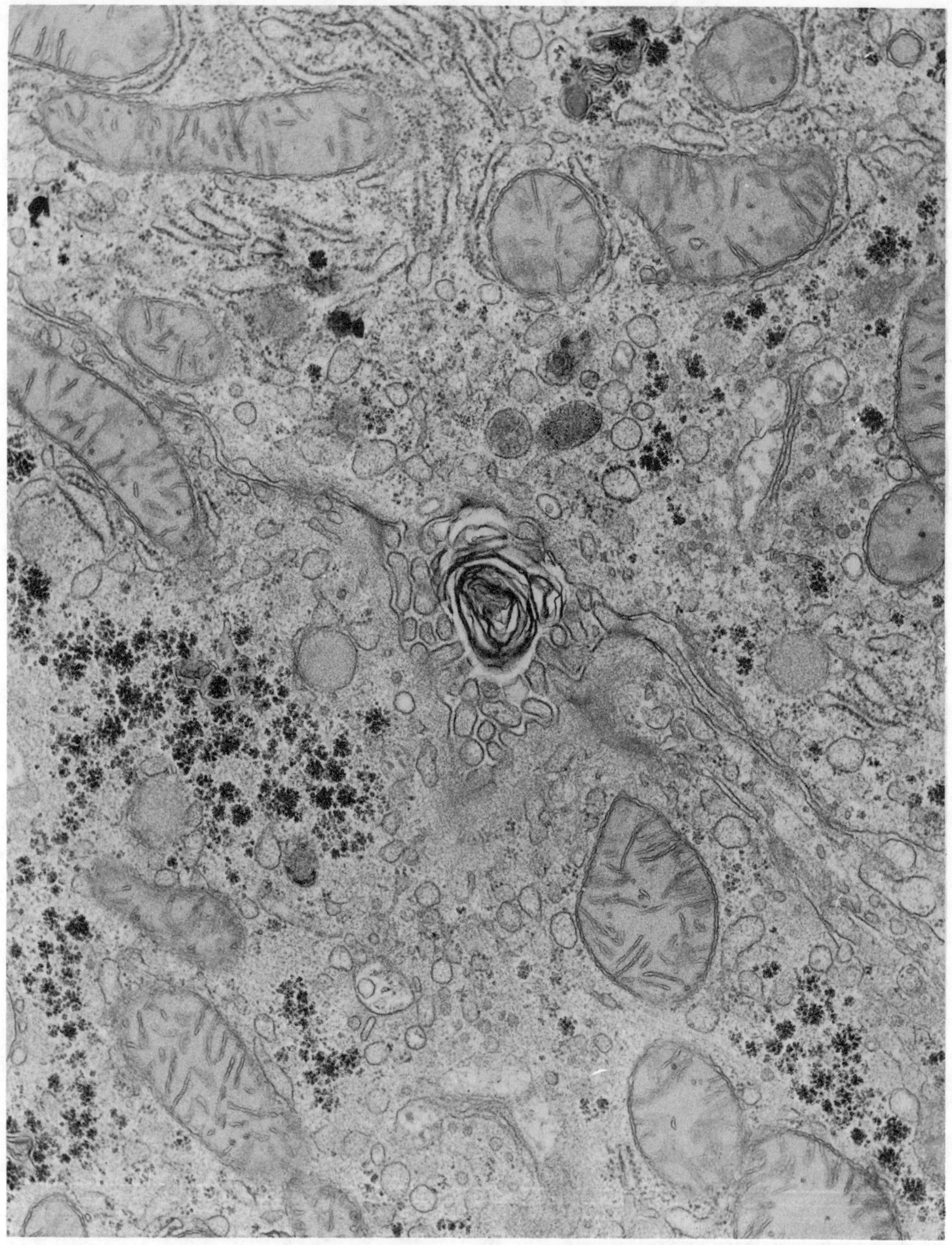

FIGURE 1.6. Electron micrograph of cytoplasm of adjacent hepatocytes with bile canaliculus between; from a mouse exposed to 500 ppm methyl chloroform, 90 min/day, 6 days/week for 1 week. Mitochondria prominent, rough and smooth endoplasmic reticulum abundant, lipid droplets are few and stain faintly with OsO_4, glycogen abundant (X 31,000).

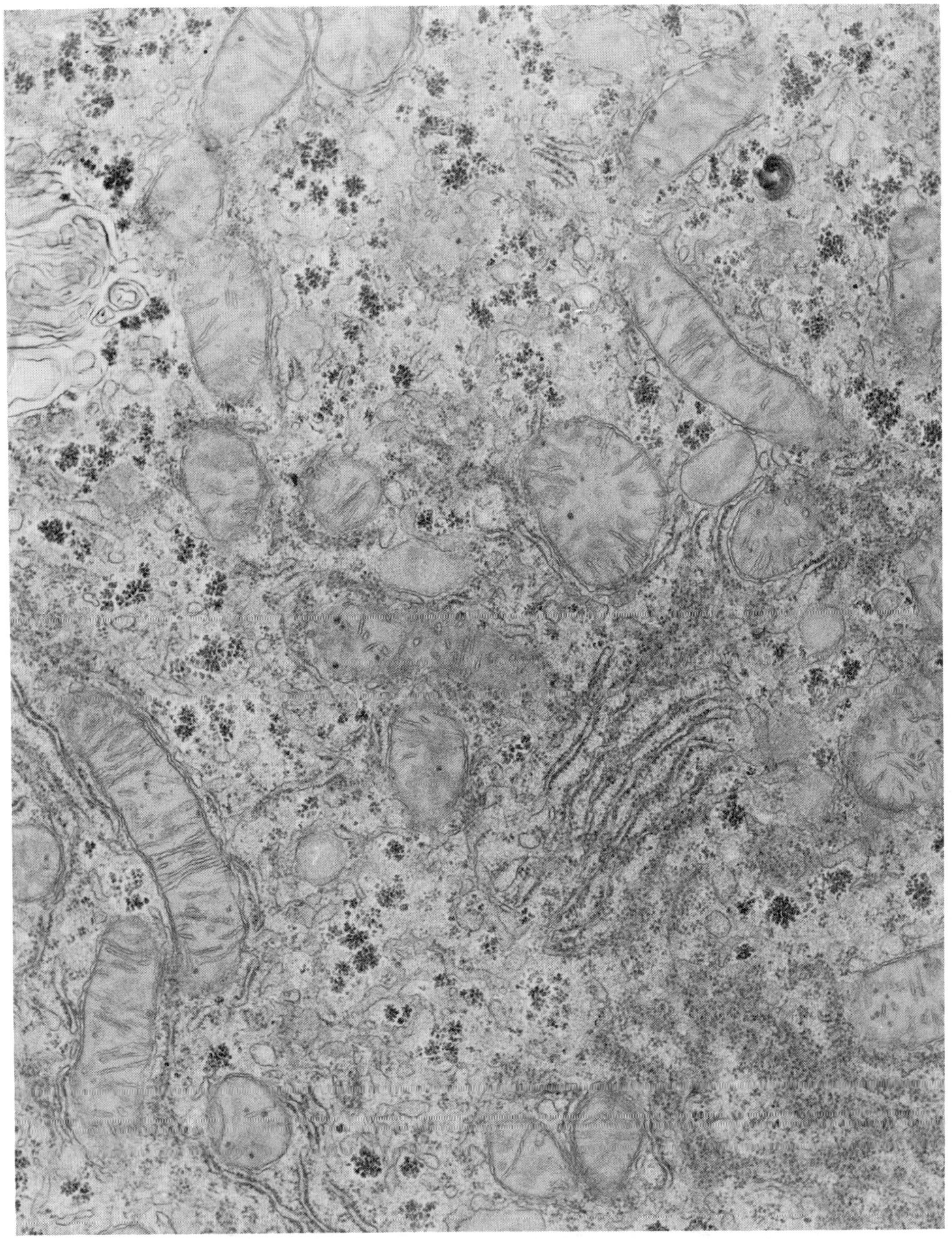

FIGURE 1.7. Electron micrograph of cytoplasm of a hepatocyte; from a mouse exposed to 2000 ppm methyl chloroform, 60 min/day, 6 days/week for 2 weeks. Mitochondria prominent, rough and smooth endoplasmic reticulum abundant, lipid droplets scanty and stain lightly with OsO_4, glycogen also abundant, membranous structures at upper left and right unidentified (× 31,000).

TABLE 1.5

Review of the Literature on Hepatotoxicity from Inhalation of Methyl Chloroform

		Exposures showing no hepatotoxicity		Exposures showing hepatotoxicity	
Species	Investigators	Concentration and duration	Hepatic measurements	Concentration and duration	Hepatic measurements
Mice	Horiguchi and Horiuchi (1971)			1000 ppm for 2 hr on alternate days; 9 exposures	Congestion of the liver and inflammation around biliary ducts[8]
	McNutt et al. (1975)	250 ppm continuously for 1–14 weeks	No increase in liver weight after 1–14 weeks except after 2 and 8 weeks; no increase in liver triglycerides except after 1, 3, 4, and 13 weeks; centrilobular balloon degeneration of hepatocytes rated in 2 out of 300 mice[9]	1000 ppm continuously for 1–14 weeks	Increase in liver to body weight ratio; increase in liver triglyceride; centrilobular balloon degeneration of hepatocytes[9]
	Lal and Shah (1970)	600 and 1500 ppm for 24 hr	No significant effect on sleeping time induced by hexobarbital[16]	3000 and 6000 ppm for 24 hr	Significant reduction of hexobarbital induced hypnosis[16]
Rats	MacEwen, Kinkead, and Haun (1974)	250 ppm continuously for 90 days	No effect in liver to body weight ratios[14]	1000 ppm continuously for 90 days	Increase in liver to body weight ratio; alterations in structure of hepatocytes[14]
	Tsapko and Rappoport (1972)			73 ppm for 4 hr daily for 50 days	Swelling of liver cells and moderately pronounced venous hyperemia[15]
	Adams, Spencer, Rowe, and Irish (1950)	5000 ppm 7 hr for 31 exposures in 44 days	No effect on liver weight and no histopathologic effects[10]	8000 ppm for 7 hr	Fatty changes in the liver; no necrosis[10]
		8000 ppm for 5 hr	No pathological findings[10]	1200 ppm for 7 hr	Increase in liver weight; fat vacuoles in hepatic cells; congestion and necrosis[10]
		18,000 ppm for 0.3 hr	No pathological findings[10]		

TABLE 1.5 (continued)

Review of the Literature on Hepatotoxicity from Inhalation of Methyl Chloroform

Species	Investigators	Exposures showing no hepatotoxicity: Concentration and duration	Exposures showing no hepatotoxicity: Hepatic measurements	Exposures showing hepatotoxicity: Concentration and duration	Exposures showing hepatotoxicity: Hepatic measurements
Rats	Torkelson, Oyen, McCollister and Rowe (1957)	500 ppm 7 hr/day, 5 days/week for 6 months	No pathological changes[12]	1000 ppm 1 hr/day, 5 days/week for 3 months	Increase in liver weight[12]
		1000 ppm for 0.05, 0.2, and 0.5 hr/day, 5 days/week for 3 months	No increase in liver weight and no pathological changes[12]		
	Prendergast, Jones, Jenkins and Siegel (1967)	135 ppm and 370 ppm continuously for 90 days	Hepatic enzymes unchanged[13]		
		2200 ppm 8 hr/day, 5 days/week for 6 weeks	Hepatic enzymes unchanged[13]		
	Fuller, Olshan, Puri and Lal (1970)	5000 ppm for 6 hr, 1000 ppm for 2, 4, and 6 hr, 1500 ppm for 2 hr	No increase in hepatic microsomal enzyme activity[17]	2500 ppm for 24 hr	Decrease duration of action of hexobarbital, meprobamate, and zoxazolamine[17]
	Carlson (1973)	11,600 ppm for 2 hr	No effect on liver weight, liver enzymes, and serum hepatic enzymes[18]	11,600 for 2 hr	Pretreatment with phenobarbital increase liver weight, increase SGPT; decreased hepatic microsomal enzyme activity[18]
Guinea pigs	Adams, Spencer, Rowe, and Irish (1950)			5000 ppm 7 hr/day, 5 days/week for 7 weeks	Slight to moderate central fatty degeneration but no necrosis[10]
	Torkelson, Oyen, McCollister, and Rowe (1958)	500 ppm 7 hr/day, 5 days/week for 6 months	No pathological changes[12]	2000 ppm, 0.5 hr/day, 5 days/week for 3 months	Fatty changes in liver[12]
		2000 ppm 0.05, 0.1 and 0.2 hr/day for 5 days/week for 3 months	No pathological changes[12]		
	Prendergast, Jones, Jenkins, and Siegel (1967)	135 and 370 ppm continuously for 90 days	No effect on serum hepatic enzymes[13]		
		2200 ppm for 8 hr/day, 5 days/week for 6 weeks	No effect on serum hepatic enzymes[13]		

TABLE 1.5 (continued)

Review of the Literature on Hepatotoxicity from Inhalation of Methyl Chloroform

Species	Investigators	Exposures showing no hepatotoxicity		Exposures showing hepatotoxicity	
		Concentration and duration	Hepatic measurements	Concentration and duration	Hepatic measurements
Rabbits	Adams, Spencer, Rowe, and Irish (1950)	5000 ppm 7 hr/day, 5 days/week for 7 weeks	No pathological changes[10]		
	Prendergast, Jones, Jenkins, and Siegel (1967)	135 and 370 ppm continuously for 90 days	No effect on serum hepatic enzymes[13]		
		2200 ppm for 8 hr/day, 5 days/week for 6 weeks	No effect on serum hepatic enzymes[13]		
	Torkelson, Oyen, McCollister, and Rowe (1958)	500 ppm 7 hr/day, 5 days/week for 6 months	No pathological changes[12]		
Dogs	Prendergast, Jones, Jenkins, and Siegel (1967)	135 and 370 ppm continuously for 90 days	No effect on serum hepatic enzymes[13]		
		2200 ppm for 8 hr/day, 5 days/week for 6 weeks	No effect on serum hepatic enzymes[13]		
	MacEwen, Kinkead, and Haun (1974)	250 and 1000 ppm continuously for 90 days	No hepatic lesions[14]		
Monkeys (*Macaca mulatta*)	Adams, Spencer, Rowe, and Irish (1950)	5000 ppm 7 hr/day, 5 days/week for 74 days	No pathological changes[10]		
	Torkelson, Oyen, McCollister, and Rowe (1958)	500 ppm 7 hr/day, 5 days/week, for 6 months	No pathological changes[12]		
	Prendergast, Jones, Jenkins, and Siegel (1967)	135 and 370 ppm continuously for 90 days	No effect on serum hepatic enzymes[13]		
		2200 ppm for 8 hr/day, 5 days/week for 6 weeks	No effect on serum hepatic enzymes[13]		
	MacEwen, Kinkead, and Haun (1974)	250 and 1000 ppm continuously for 90 days	No hepatic lesions[14]		

TABLE 1.5 (continued)

Review of the Literature on Hepatotoxicity from Inhalation of Methyl Chloroform

Species	Investigators	Exposures showing no hepatotoxicity: Concentration and duration	Exposures showing no hepatotoxicity: Hepatic measurements	Exposures showing hepatotoxicity: Concentration and duration	Exposures showing hepatotoxicity: Hepatic measurements
Human subjects	Torkelson, Oyen, McCollister, and Rowe (1958)	506 ppm for 7.5 hr	No effect on urinary urobilirubinogen, thymol turbidity and excretion of bromsulfophthalein[12]		
		920 ppm for 1.2 hr	No effect on urinary urobilirubinogen, thymol turbidity and excretion of bromsulfophthalein[12]		
	Stewart, Gay, Erley, Hale, and Schaffer (1961)	900 ppm for 0.3 hr	No effect on urinary urobilirubinogen in 2 of 3 subjects; no effect on serum hepatic enzymes in all 3 subjects[11]	900 ppm for 0.3 hr	Elevated urinary urobilirubinogen in 1 of 3 subjects[11]
		0 to 2650 ppm for 0.25 hr	No effect on urinary urobilirubinogen in 5 of 7 subjects; no effect on serum hepatic enzymes in all 7 subjects[11]	0 to 2650 ppm for 0.25 hr	Elevated urinary urobilirubinogen in 2 of 7 subjects[11]
		500 ppm for 1.2 hr	No effect on serum hepatic enzymes or on urinary urobilirubinogen in all subjects[11]		
	Stewart, Gay, Erley, Hale, and Schaffer (1961)	500 ppm for 3 hr	No effect on serum hepatic enzymes or on urinary urobilirubinogen in all subjects[11]		
	Stewart, Gay, Schaffer, Erley, and Rowe (1969)	500 ppm for 6.5 to 7 hr/day for 5 days	No effect on serum hepatic enzymes[19]		

TABLE 1.5 (continued)

Review of the Literature on Hepatotoxicity from Inhalation of Methyl Chloroform

Species	Investigators	Exposures showing no hepatotoxicity		Exposures showing hepatotoxicity	
		Concentration and duration	Hepatic measurements	Concentration and duration	Hepatic measurements
	Dornette and Jones (1969)	Anesthesia up to 2 hr	No effect on serum hepatic enzymes in 3 of 5 patients[20]	Anesthesia up to 2 hr	Increase in serum hepatic enzymes in 2 of 5 patients[20]
	Seki, Yoshwka, Aikawa, Maleumura, Ichikawa, Hiratsuka, Yoshioka, Shimbo, Ikeda (1975)	4, 25, 28, and 53 ppm occupational exposure	No effect on urinary urobilirubinogen in 196 workers in 4 plants[21]		
	Weitbrecht (1965)			10–40 ppm occupational exposure	Increase in urinary urobilirubinogen in 2 of 9 workers[22]

Chapter 2
METHYLENE CHLORIDE

The introduction of halogens into the molecule of hydrocarbons exerts a pronounced effect not only on the physicochemical properties, but also on the biological behavior of these organic compounds. Halogen substitution usually intensifies the pharmacologic as well as toxicologic properties of a given compound. In a previous study, the acute[1] and subacute toxicity (Chapter 1) of a trihalogen compound, namely methyl chloroform (1,1,1-trichloroethane), was investigated. A similar monograph on a dihalogen compound, methylene chloride (or dichloromethane), will be published as part of this series on solvents. However, the continued use of methylene chloride as a solvent for aerosols and for paint removers has prompted a partial presentation prior to the appearance of the monograph. The objective of the present chapter is to examine and compare the acute toxicity in mice and dogs of inhibited methylene chloride (Aerothene MM®, the Dow Chemical Company), methyl chloroform, and a commercially available paint remover. The paint remover used in this study consists of 90.2% w/w methylene chloride, 4.2% w/w methanol, 3.2% w/w isopropanol, and 2.4% w/w of toluene.

A. Methods

1. Acute Toxicity in Mice

The lethal dose (concentration) for 50% of the animals (LD_{50} and LC_{50}) was determined in male mice (CF-1 strain, Charles River Laboratories, Wilmington, Massachusetts). Mice weighing from 20 to 25 g were used and the mean weight of mice was approximately the same in all groups receiving the same compound via the same route of administration. In the determination of the LC_{50}, groups of ten mice each were introduced into a 10-liter glass chamber and various concentrations of the solvent were flushed into the chamber at a rate of 2 to 3 l/min for 20 min. Mortality rate was determined and LD_{50} and LC_{50} were calculated according to the probit method.

2. Acute Toxicity in Dogs

Adult mongrel dogs of either sex, weighing between 16 and 22 kg (average 18 kg) were anesthetized with 30 to 35 mg/kg pentobarbital sodium injected intravenously. Dogs were artificially ventilated with room air via an endotracheal cannula and the aortic blood pressure was measured by inserting a catheter connected to a pressure transducer (Statham P23AA) into the right carotid artery. Thoracotomy was then carried out on the left side of the animal at the fourth intercostal space. The lungs were gently retracted and the pericardium was opened. The pulmonary artery was separated from the aorta and an electromagnetic flow probe with an internal diameter of approximately 10% less than the pulmonary artery was fitted around the latter shortly before its bifurcation. The pulmonary arterial flow was thus monitored using a Statham SP 2202 electromagnetic flowmeter. Left atrial pressure was measured by introducing a catheter, through a small slit, in the left atrial appendage, connected to a P23BB Statham pressure transducer. Likewise, pulmonary arterial pressure was measured by inserting a catheter into the pulmonary artery to the left lower lobe. To measure the left ventricular pressure, the chest was opened at the sixth intercostal space of the right side and a catheter was inserted into the left ventricular lumen via the myocardial apex. The signal was fed to a differentiating circuit (8814 A Hewlett Packard derivative computer) and the left ventricular dp/dt was recorded. All parameters were recorded on a six-channeled Sanborn 7700 recorder. Figure 2.1 illustrates the connections to the heart and blood vessels for measuring the hemodynamic effects of methylene chloride.

Hemodynamic effects of methylene chloride – The preparation was allowed to stabilize for a period of up to 30 min, after which the following concentrations of methylene chloride (0.5%, 1.0%, 2.5%, and 5% v/v) in air were administered via the inlet of the respirator for 5 min. Each subsequent concentration was administered to the animal immediately after the preceding one. Various concentrations of methylene chloride were prepared by volatilizing the appropriate amount of the solvent into a known volume of air as previously described.[1]

In studies of the open-chest dogs before and after total alpha- and beta-adrenergic blockade, administration of 1% concentration of methylene chloride was carried out before and after injecting 0.1 mg/kg phentolamine mesylate followed after 10 min by 0.5 mg/kg propranolol hydrochloride

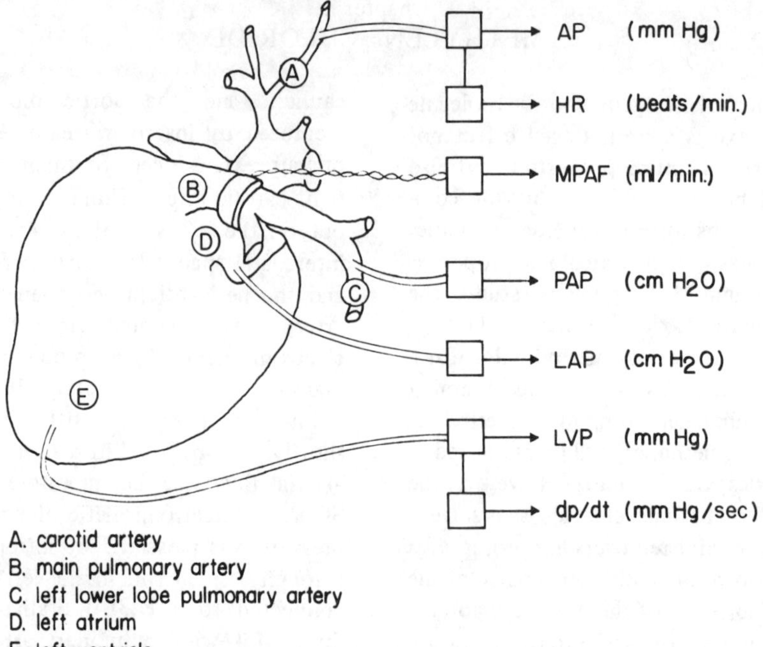

FIGURE 2.1. Illustration of the various myocardial and hemodynamic parameters measured in the experiments.

intravenously. The arrhythmogenic effect of 5% methylene chloride and 2.5% methyl chloroform were also studied in a similar preparation.

All data were analyzed by paired comparison. A Student t-test was used to analyze the effect before and after the administration of adrenergic blocking agents. In both cases, the criterion for significance was P less than or equal to 0.05.

Hemodynamic effects of paint remover – After the preparation was stabilized, the appropriate volumes of paint remover were totally volatilized in a suitable volume of air and were administered via the inlet of respiratory valve for 5 min according to the following sequence of increasing concentrations (v/v): 0.5%, 1.0%, 2.5%, and 5.0%.

The paired t-test was used to determine the statistical significance of mean differences between control and experimental values. A Student t-test was used to test the significance of the differences between the effects of methylene chloride and those of the paint remover. In both cases, differences were considered statistically significant if P values were equal to or less than 0.05.

3. Calculations and Abbreviations of Circulatory Measurements

MPAP: Mean pulmonary arterial pressure in cm H_2O, measured from the pulmonary artery to the left lower lobe.

MLAP: Mean left atrial pressue in cm H_2O, measured from the left atrium.

EMPAP: Effective mean pulmonary arterial pressure in cm H_2O, mean pulmonary arterial pressure minus mean left atrial pressure.

LVP: Left ventricular pressure in mmHg, measured from a catheter in the left ventricular cavity inserted through the cardiac apex.

LVEDP: Left ventricular end-diastolic pressure in mmHg, measured from the left ventricular pressure.

dp/dt: Maximal rate of rise of left ventricular pressure in mmHg/sec, derived from the left ventricular pressure with a derivative computer, type HP 8814A.

MAP: Mean aortic pressure in mmHg, measured from a catheter inserted through the carotid artery.

MPAF: Mean pulmonary arterial flow in ml/min, measured with a Statham electromagnetic flow probe around the main pulmonary artery.

HR: Heart rate, in beats/min, computed from aortic pressure waves.

Systemic vascular resistance: In dynes·sec/cm^5; the quotient of mean aortic pressure minus left ventricular end-diastolic pressure in dynes/cm^2 and of mean pulmonary arterial flow in ml/sec.

Pulmonary vascular resistance: In dynes·sec/cm^5; the quotient of mean pulmonary arterial pressure minus mean left atrial pressure in dynes/cm^2 and of mean pulmonary arterial flow in ml/sec.

Stroke volume: In ml, computed from cardiac output and heart rate.

Stroke work: In gram·meter computed from stroke volume and aortic pressure.

B. Results

1. Acute Toxicity in Mice

Data obtained in acute toxicity experiments where methylene chloride was given orally, intraperitoneally, or by inhalation are summarized in Tables 2.1 to 2.3. The corresponding data for the paint remover are shown in Tables 2.4 to 2.6. A comparison of the LD_{50} and LC_{50} of both solvents is presented in Table 2.7.

The oral LD_{50} values of methylene chloride and paint remover are 1.987 and 1.997 mg/kg, respectively. The intraperitoneal LD_{50} values are 448 mg/kg and 758 mg/kg for methylene chloride and paint remover, respectively. The corresponding inhalational LC_{50} (20 min) values are 2.67% and 3.57%.

Mice receiving either methylene chloride or paint remover showed signs of irritation and restlessness, Straub tail, ataxia, and rapid and shallow respiration followed by death.

2. Hemodynamic Effects of Methylene Chloride in Dogs

A record of a typical experiment consisting of the inhalation of 0.5, 1.0, 2.5, and 5.0% of methylene chloride is depicted in Figure 2.2. The hemodynamic effects representing the mean and standard error of results from five dogs are summarized in Table 2.8. The administration of various concentrations of methylene chloride in the intact open-chest dog preparation brought about a biphasic pattern of response. Thus, at lower concentrations (0.5 and 1.0%), increases in left ventricular pressure, left ventricular dp/dt, mean arterial pressure, and stroke work were observed. Decreases in the same parameters were noticed after the administration of higher concentrations (2.5% and 5%) of methylene chloride.

Administration of the threshold concentration of 0.5% of methylene chloride brought about a mean increase of 2.3%, 1.2%, 2.7%, and 2.8% in left ventricular pressure, left ventricular dp/dt, mean arterial pressure, and stroke work, respectively. Inhalation of twice the threshold concentration of methylene chloride (1.0%) brought about an increase of 3.1%, 0.7%, and 4.3%, and a decrease of 0.2% in the above-mentioned parameters, respectively. A decrease in effective mean pulmonary arterial pressure of 2.6% and 4.3%, and a decrease in pulmonary vascular resistance averaging to 2.1 and 1.9% of the control value, were observed after inhalation of 0.5% and 1% concentrations of methylene chloride, respectively. The change in effective mean pulmonary arterial pressure was concentration-dependent and a significant decrease averaging to 7.3% and 29.1% was observed after inhalation of 2.5% and 5% methylene chloride, respectively. Nonetheless, these higher concentrations, viz., 2.5% and 5%, brought about an increase in pulmonary vascular resistance of 3.0% and 8.9%, respectively. A concentration-dependent decrease in left ventricular pressure of 0.9% and 3.9% in left ventricular dp/dt of 7.5% and 20.9%, and in stroke work of 8.3% and 35.5%, was observed after administration of 2.5% and 5.0% methylene chloride, respectively. A consistent decrease in mean pulmonary arterial flow of 0.2%, 2.0%, 10.0%, and 35.3% of the control value, and a decrease in stroke volume amounting to 0.1%, 4.6%, 11.1%, and 31.4% of the control value, were observed after inhalation of 0.5%, 1.0%, 2.5%, and 5.0% of methylene chloride, respectively.

Influence of phentolamine and propranolol – Figure 2.3 shows the various hemodynamic changes brought about by 1% methylene chloride. Increases in left ventricular pressure, left ventricular dp/dt, heart rate, mean arterial pressure, mean pulmonary arterial pressure, and systemic vascular resistance were observed after inhalation of 1% methylene chloride in the open-chest dog preparation. Decreases in mean pulmonary arterial flow, effective mean pulmonary arterial pressure, stroke volume, and stroke work were also observed after the inhalation of 1% methylene chloride. The only significant change obtained after treatment with phentolamine mesylate and inhalation of 1% methylene chloride was a decrease in mean pulmonary arterial pressure. However, administration of the same concentration of methylene chloride after blocking beta-adrenergic receptors

TABLE 2.1

Acute Toxicity of Methylene Chloride Given Orally to Mice

Group	Dose g/kg	Number of mice	% Mortality	LD_{50} (g/kg) 24 hr	Regression coefficient
1	0.65	10	0		
2	1.33	9	22.2		
3	1.95	11	45.6	1.987 ± 0.398	0.9574
4	2.64	11	72.7		
5	3.82	12	91.7		
6	5.10	12	100		

TABLE 2.2

Acute Toxicity of Methylene Chloride Given Intraperitoneally to Mice

Group	Dose mg/kg	Number of mice	% Mortality	LD_{50} (mg/kg) 24 hr	Regression coefficient
1	113.7	11	0		
2	230.5	12	25		
3	338.4	10	40	448.11 ± 73.45	0.9472
4	457.0	11	63.6		
5	566.4	12	83.3		
6	1132.9	12	100		

TABLE 2.3

Acute Inhalational Toxicity of Methylene Chloride in Mice After 20 min of Exposure*

Group	Concentration % v/v	Concentration mg/l	% Mortality	LC_{50}	Regression coefficient
1	1	34.74	0		
2	2	69.48	20		
3	3	104.22	60	2.671 ± 0.454%	0.9845
4	4	138.96	90		
5	5	173.70	100		

*Each group consisted of ten mice.

TABLE 2.4

Acute Toxicity of Paint Remover Given Orally to Mice

Group	Dose g/kg	Number of mice	% Mortality	LD_{50} (g/kg) 24 hr	Regression coefficient
1	0.753	9	0		
2	1.418	12	25		
3	2.173	11	54.5	1.997 ± 0.378	0.9778
4	2.962	10	80		
5	3.582	11	90.9		
6	4.418	10	100		

TABLE 2.5

Acute Toxicity of Paint Remover Given Intraperitoneally to Mice

Group	Dose mg/kg	Number of mice	% Mortality	LD_{50} (mg/kg) 24 hr	Regression coefficient
1	113.2	12	0		
2	283.2	12	16.7		
3	571.3	12	41.7	758.38 ± 265.17	0.9799
4	1137.7	12	50		
5	1728.9	12	83.3		
6	2295.2	12	100		

TABLE 2.6

Acute Inhalational Toxicity of Paint Remover in Mice After 20 min of Exposure

Group*	Concentration % v/v	% Mortality	LC_{50}	Regression coefficient
1	2	0		
2	3	30		
3	4	70	3.565 ± 0.473%	0.9790
4	5	80		
5	6	100		

*Each group consisted of ten mice.

TABLE 2.7

Comparison of LD_{50} and LC_{50} of Methylene Chloride and a Paint Remover (90.17% Methylene Chloride) in Mice, Using Various Routes of Administration

	LD_{50}		LC_{50}
Compound	Oral g/kg	Intraperitoneal mg/kg	Inhalation (20 min) % v/v
Methylene chloride (MCL)	1.987	448.11	2.671
Paint remover (PR)	1.997	758.38	3.565
Toxicity of PR in relation to MCL: $\frac{PR}{MCL}\%$	99.5	59.1	74.9

with propranolol hydrochloride elicited significant decreases in left ventricular pressure, left ventricular dp/dt, heart rate, mean arterial pressure, mean pulmonary arterial flow, effective mean pulmonary arterial pressure, stroke volume, and stroke work. A record of a typical experiment is shown in Figure 2.4.

Comparison of methyl chloroform and methylene chloride – The concentration-response curves for methyl chloroform and methylene chloride for various parameters are depicted in Figure 2.5 and Figure 2.6. Generally, methyl chloroform is effective in lower concentrations than methylene chloride; thus, while the threshold

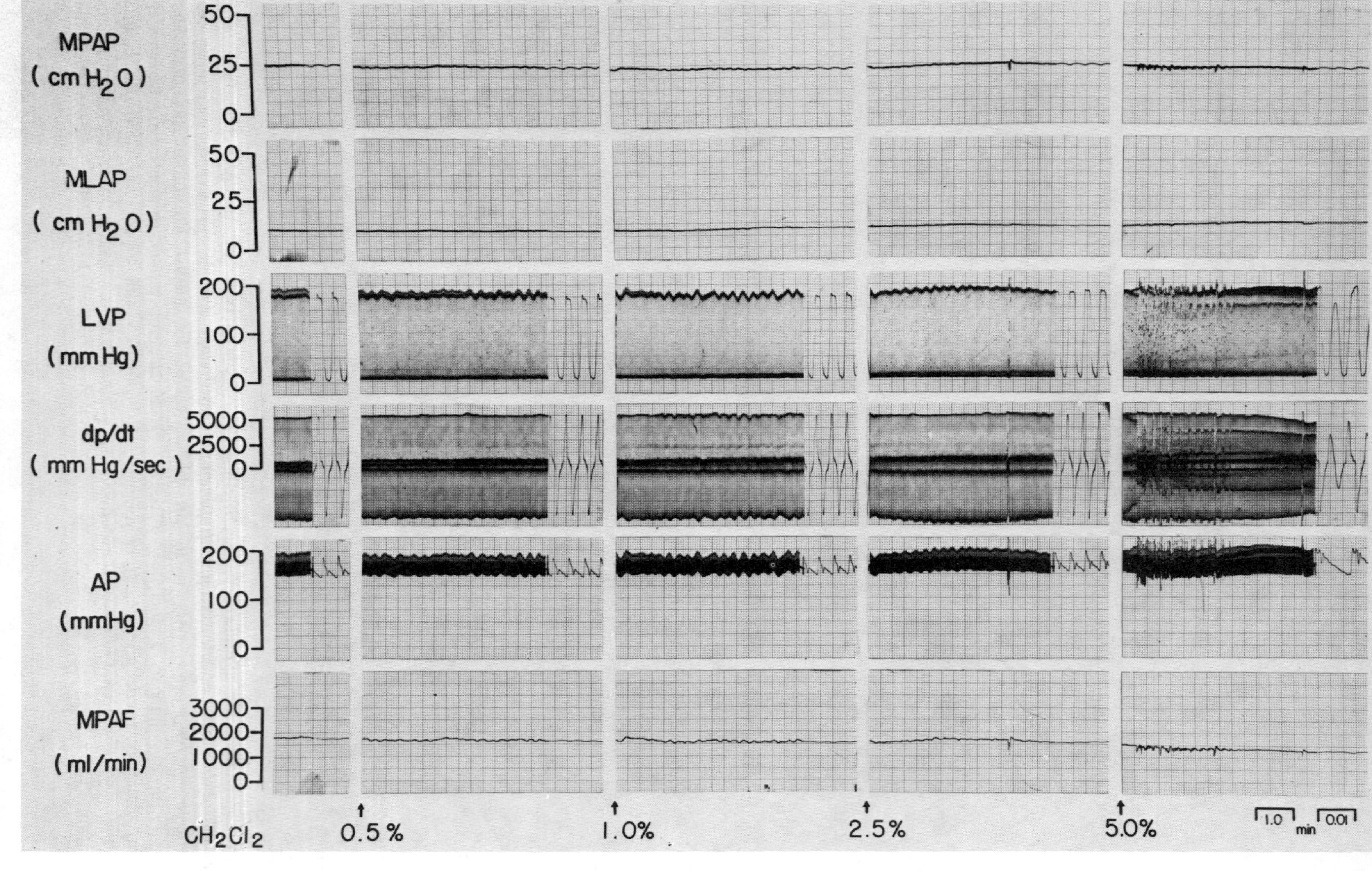

FIGURE 2.2. The effect of progressively increasing concentrations of methylene chloride administered by inhalation to the open-chest dog preparation. Note the persistent arrhythmia induced by 5%.

TABLE 2.8

The Effects of Progressively Increasing Concentrations of Methylene Chloride on the Hemodynamics of the Anesthetized Open-chest Dog Preparation*

	MPAP cm H_2O	MLAP cm H_2O	EMPAP cm H_2O	LVP mmHg	LVEDP mmHg	dp/dt mmHg/sec	MAP mmHg	MPAF ml/min	HR beats/min	Vascular resistance dynes·sec/cm^5 Pulmonary	Vascular resistance dynes·sec/cm^5 Systemic	Stroke vol ml	Stroke work g·meter
Control	37.6 ± 3.3	7.6 ± 1.2	31.5 ± 2.5	151 ± 14.3	4 ± 1.0	4150 ± 322	130 ± 12.3	1550 ± 125	191 ± 11	1218 ± 113	6592 ± 686	8.2 ± 0.7	15.1 ± 1.9
0.5%	37.2 ± 2.9	8.2 ± 1.2	30.5 ± 2.2	153 ± 12.2	3.6 ± 1.0	4200 ± 332	133 ± 10.9	1540 ± 109	191 ± 10	1196 ± 127	6798 ± 672	8.2 ± 0.7	15.4 ± 1.9
	−0.4 ± 0.7	+0.5 ± 0.4	−0.9 ± 0.5	+2.6 ± 2.6	−0.4 ± 1.0	+50 ± 30.6	+3 ± 1.9	−10 ± 40	−0.8 ± 1.2	−22 ± 29	206 ± 112	−0.1 ± 0.3	+0.3 ± 0.5
	NS	NS	NS	NS	NS	NS	NS	NS	NS	NS	NS	NS	NS
1.0%	37.1 ± 2.9	8.5 ± 1.1	30.1 ± 2.3	154 ± 11.2	4 ± 1.0	4225 ± 462	134 ± 10.0	1510 ± 107	196 ± 11	1203 ± 136	6989 ± 615	7.8 ± 0.7	14.8 ± 1.5
	−0.5 ± 0.7	+0.9 ± 0.4	−4.1 ± 0.6	+3.4 ± 3.7	0 ± 0	+75 ± 179.4	+4 ± 3.0	−40 ± 53	+5 ± 2.4	−15 ± 34	+396 ± 187	−0.4 ± 0.3	−0.3 ± 0.8
	NS	NS	NS	NS	NS	NS	NS	NS	NS	NS	NS	NS	NS
2.5%	37.1 ± 3.0	9.4 ± 1.0	29.2 ± 2.5	148 ± 9.8	5 ± 1.6	3850 ± 370	133 ± 10.0	1390 ± 101	193 ± 11	1260 ± 135	7422 ± 563	7.2 ± 0.6	13.7 ± 1.5
	−0.5 ± 0.8	−1.8 ± 0.8	−2.3 ± 0.7	−3.0 ± 4.7	+1 ± 0.5	−300 ± 145.8	+3 ± 3.9	−160 ± 37	+1.8 ± 1.2	+42 ± 26	+830 ± 259	+0.9 ± 0.2	− 1.4 ± 0.6
	NS	NS	0.05	NS	NS	NS	NS	0.01	NS	NS	0.05	0.01	NS
5.0%	36.0 ± 2.4	14.2 ± 1.9	22.6 ± 3.4	144 ± 13.0	9 ± 1.9	3225 ± 384	125 ± 11.7	990 ± 93	170 ± 7	1340 ± 175	9508 ± 936	5.9 ± 0.4	8.6 ± 0.1
	−1.6 ± 1.9	+6.5 ± 2.2	−8.9 ± 2.1	−6.6 ± 6.2	+5.0 ± 3.2	−925 ± 441	−5 ± 3.7	−560 ± 117	−4.5 ± 7.5	+122 ± 88	+2915 ± 577	−2.7 ± 0.2	−4.2 ± 0.4
	NS	0.05	0.02	NS	0.05	NS	NS	0.01	NS	NS	0.01	0.001	0.001

*Each group of numbers includes mean ± standard error for five dogs, mean difference ± standard error, and the significance level. NS = nonsignificant. Abbreviations are explained in the text.

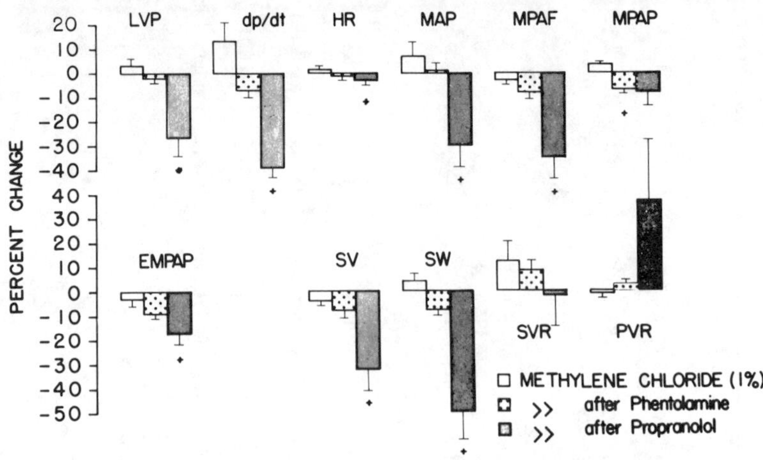

FIGURE 2.3. Percent change in various measured (upper) and derived (lower) parameters before and after total adrenergic blockade. Asterisk indicates significant change from the effect of 1% methylene chloride before blockade of adrenergic receptors.

effective concentration of the former is 0.1%, it is 0.5% in the latter. Except for the initial phase of increase in contractility following methylene chloride, both compounds share the property of depressing the myocardium as indicated from the decrease of the rate of rise of left ventricular dp/dt. At higher concentrations, both compounds bring about a decrease in mean pulmonary arterial flow, left ventricular pressure, and mean arterial pressure. However, while the most prominent effect of methyl chloroform is a decrease in myocardial contractile force, that of methylene chloride is a decrease in cardiac output as reflected from the decrease in mean pulmonary arterial flow. Furthermore, plotting of the calculated values for systemic and pulmonary vascular resistances (Figure 2.6) revealed a striking difference between both compounds. Thus while methyl chloroform brought about a decrease in systemic vascular resistance and a pronounced increase in pulmonary vascular resistance, methylene chloride, at higher concentrations, elicited a moderate increase in pulmonary vascular resistance. On the other hand, both compounds showed the ability to induce myocardial arrhythmia in the dog. The concentrations that brought about persistent arrhythmia are 2.5% for methyl chloroform and 5.0% for methylene chloride. A record of a typical experiment is shown in Figure 2.7.

3. Hemodynamic Effects of Paint Remover in Dogs

Inhalation of various concentrations of paint remover was associated with various myocardial and hemodynamic changes. The results are summarized in Table 2.9 and the record of a representative experiment is shown in Figure 2.8. The first significant change observed was a small increase in systemic arterial pressure. This effect was induced by as low a concentration as 0.5% and amounts to 2.1% of the average control. Increasing the inhaled concentration to 1.0%, 2.5%, and 5% intensified this response and was associated with an increase of 4.1%, 6.8%, and 7.9%, respectively. The corresponding changes in left ventricular pressure, after inhalation of concentrations of 0.5%, 1.0%, 2.5%, and 5.0% were +1.6%, +0.8%, +2.5%, and +2.1%, respectively.

The inhalation of paint remover caused an initial increase followed by a decrease in myocardial contractility and cardiac output. A nonsignificant increase in myocardial contractility, as gauged by the maximal rate of rise of left ventricular dp/dt, averaging to 0.4% of the control value, was observed after inhalation of 0.5%. Increasing the concentration of paint remover in the inhaled mixture to 1.0% and 2.5% decreased myocardial contractility by 1.2% and 6.0%, respectively. Increasing the concentration of paint remover to 5.0% brought about a decrease in myocardial contractility averaging to 24.5% of the control average. A nonsignificant increase in mean pulmonary arterial flow averaging to 0.3% and 0.8% was observed after inhalation of 0.5% and 1.0%, respectively. Increase in the concentration of the inhaled mixture to 2.5% and 5.0%, however, decreased cardiac output by 0.9% and 9.2%,

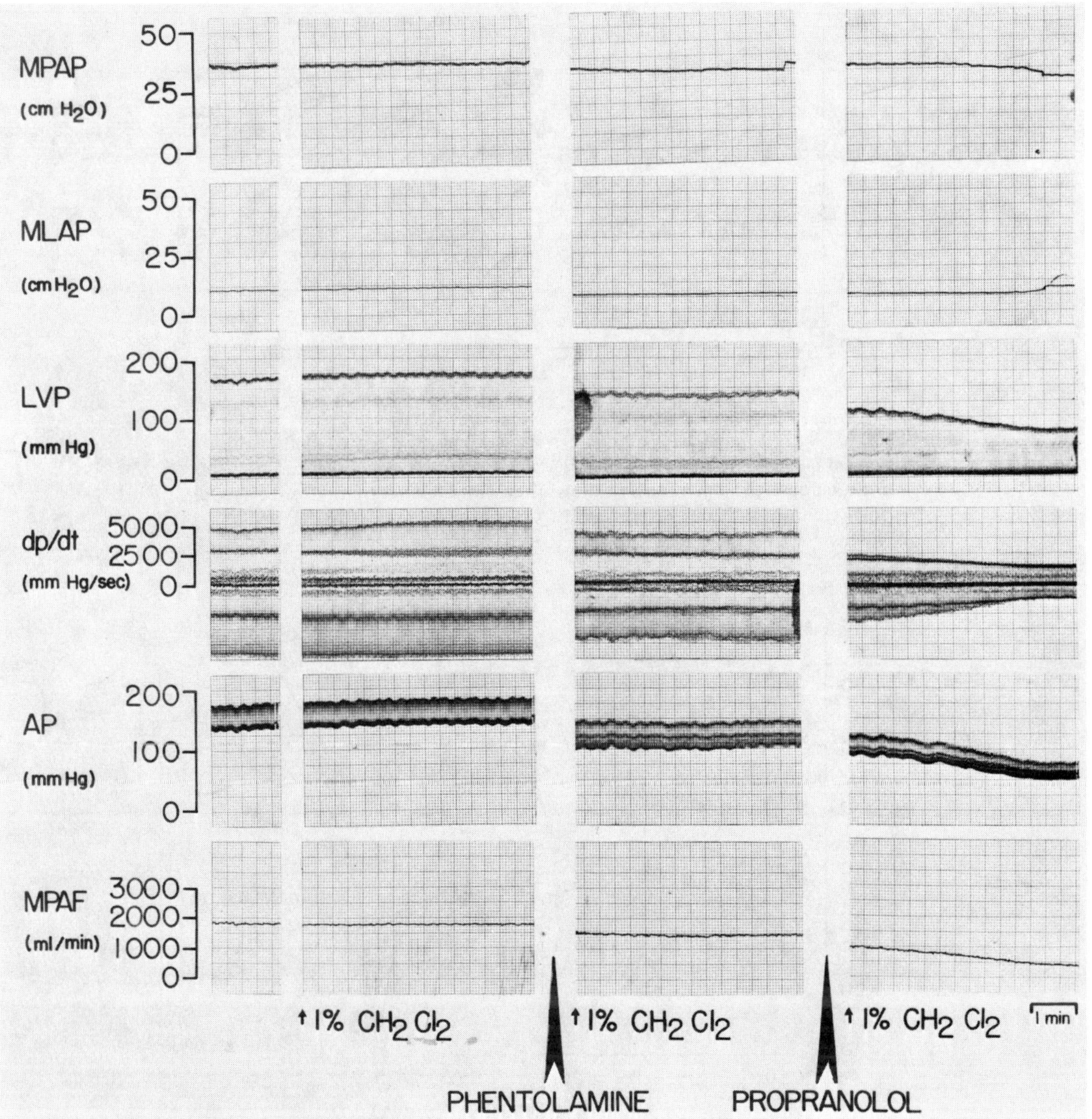

FIGURE 2.4. The effect of 1.0% v/v concentration of methylene chloride administered by inhalation to the open-chest dog before and after total alpha- and beta-adrenergic blockade. Note the decrease in all parameters except left atrial pressure.

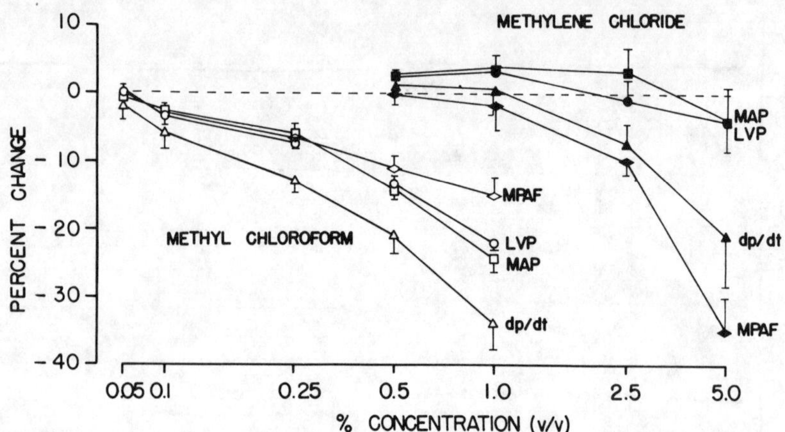

FIGURE 2.5. Comparative effects of various concentrations of methyl chloroform and methylene chloride on the arterial pressure, left ventricular pressure, left ventricular dp/dt, and pulmonary arterial flow. Note the most prominent effect of methyl chloroform is on the myocardial force of contraction (dp/dt), while that of methylene chloride is on the cardiac output.

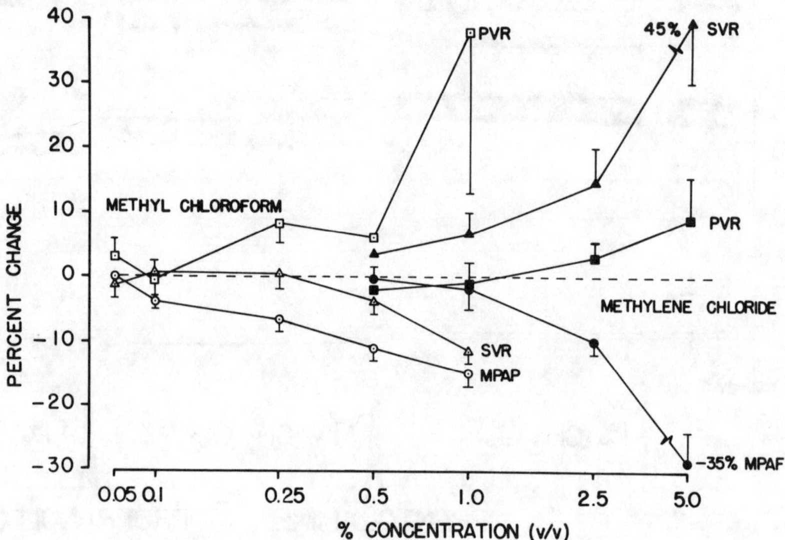

FIGURE 2.6. The effect of various concentrations of methyl chloroform and methylene chloride on the systemic and pulmonary vascular resistances. Note that while methyl chloroform decreased systemic vascular resistance, methylene chloride increased the same parameter.

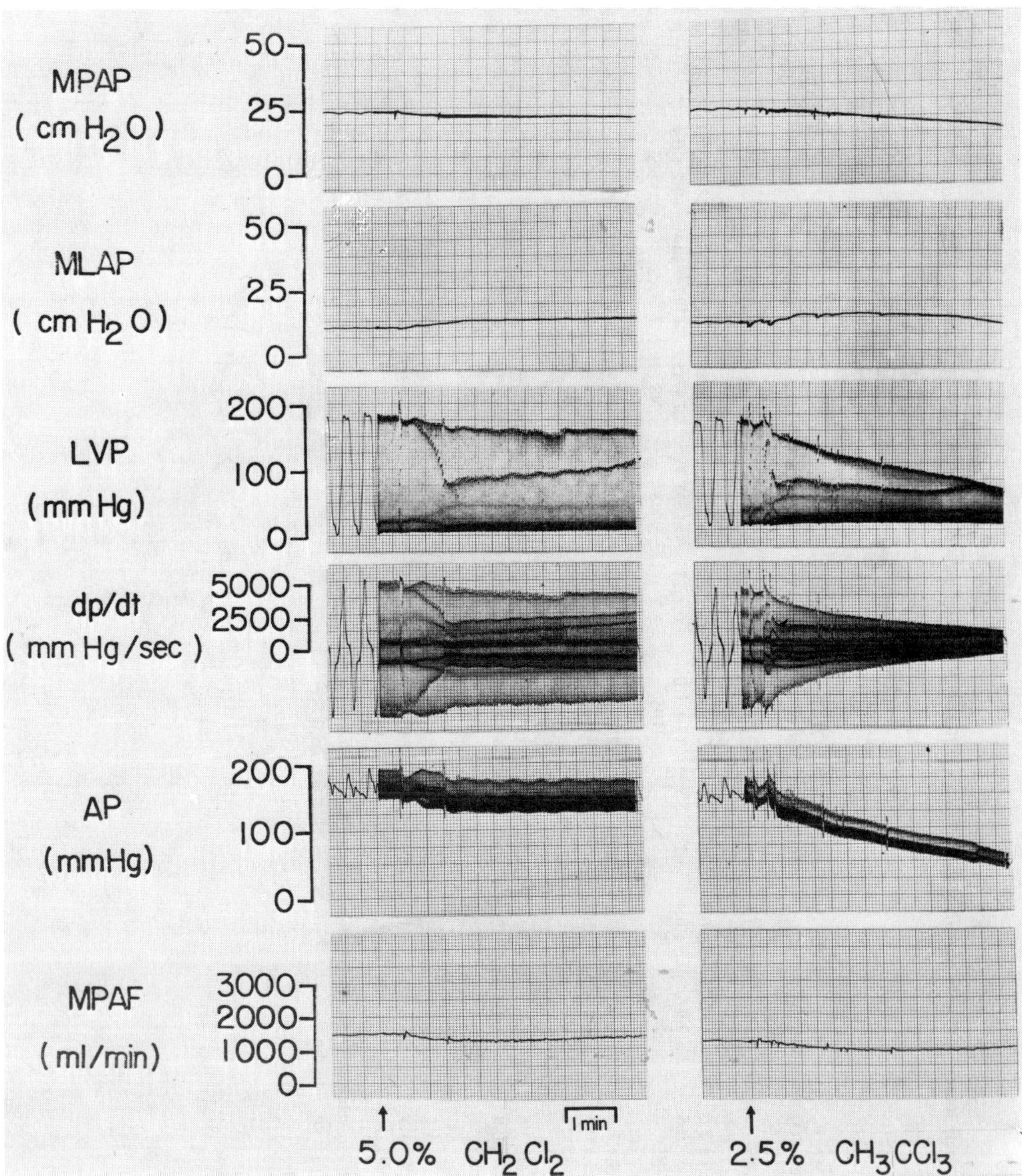

FIGURE 2.7. The arrhythmogenic effect of 2.5% methyl chloroform and 5.0% methylene chloride on the open-chest dog preparation. Note the large decrease in left ventricular pressure, left ventricular dp/dt, aortic pressure, and pulmonary arterial flow under the influence of methyl chloroform as compared with the qualitatively similar but less pronounced effect of methylene chloride.

TABLE 2.9

The Effects of Progressively Increasing Concentrations of Paint Remover (PR) on the Hemodynamics of Anesthetized Open-chest Dog Preparation*

	MPAP cm H_2O	MLAP cm H_2O	EMPAP cm H_2O	LVP mmHg	LVEDP mmHg	dp/dt mmHg/sec	MAP mmHg	MPAF ml/min	HR beats/min	Vascular resistance dynes·sec/cm^5 Pulmonary	Vascular resistance dynes·sec/cm^5 Systemic	Stroke vol ml	Stroke work g·meter
Control	27.3 ± 1.7	6.5 ± 1.4	20.9 ± 1.8	145 ± 10	4.6 ± 1.0	3196 ± 149	122 ± 4	1483 ± 70	177 ± 8	832 ± 80	6313 ± 253	8.2 ± 0.9	14.3 ± 0.9
PR 0.5%	27.4 ± 2.2	6.7 ± 1.4	20.7 ± 2.0	147 ± 10	4.3 ± 0.9	3214 ± 171	125 ± 4	1483 ± 52	181 ± 7	825 ± 86	6440 ± 253	8.0 ± 0.3	14.3 ± 0.8
	+0.1 ± 0.6	+0.2 ± 0.2	−0.2 ± 0.5	+2 ± 0.9	−0.3 ± 0.4	+18 ± 57	+3 ± 1.0	0 ± 22	+4 ± 1.9	−7 ± 15	+127 ± 114	−0.2 ± 0.1	0 ± 0.3
	NS	NS	NS	NS	NS	NS	0.05	NS	NS	NS	NS	NS	NS
PR 1.0%	28.2 ± 2.5	6.8 ± 1.4	21.4 ± 1.9	146 ± 10	5.1 ± 1.1	3143 ± 106	127 ± 4	1492 ± 55	184 ± 8	849 ± 90	6542 ± 182	7.9 ± 0.3	14.5 ± 0.9
	+0.9 ± 1.1	+0.3 ± 0.3	+0.5 ± 0.9	+1 ± 2.2	+0.5 ± 0.7	−53 ± 86	+5 ± 1.3	+9 ± 27	+7 ± 2.6	+17 ± 22	+229 ± 135	−0.3 ± 0.2	+0.2 ± 0.4
	NS	NS	NS	NS	NS	NS	0.01	NS	0.05	NS	NS	NS	NS
PR 2.5%	28.2 ± 2.7	7.6 ± 1.4	20.6 ± 2.0	148 ± 8	6.1 ± 1.4	2982 ± 74	130 ± 4	1467 ± 56	182 ± 7	836 ± 99	6702 ± 262	7.8 ± 0.3	14.6 ± 1.0
	+0.9 ± 1.2	+1.1 ± 0.5	−0.3 ± 0.9	+3 ± 2.9	+1.5 ± 0.7	−214 ± 118	+8 ± 1.7	−16 ± 16	+5 ± 2.5	+4 ± 30	+389 ± 164	−0.4 ± 0.1	+0.3 ± 0.4
	NS	NS	NS	NS	NS	NS	0.01	NS	NS	NS	NS	0.05	NS
PR 5.0%	28.8 ± 2.5	11.7 ± 1.6	17.0 ± 1.2	146 ± 6	8.2 ± 1.8	2393 ± 143	131 ± 4	1350 ± 85	170 ± 7	761 ± 85	7230 ± 327	7.6 ± 0.5	14.4 ± 1.2
	+1.5 ± 1.3	+5.2 ± 1.4	−3.9 ± 1.9	+1 ± 5.2	+3.6 ± 1.2	−803 ± 189	+9 ± 2.9	−133 ± 36	−7 ± 5.0	−71 ± 60	+917 ± 351	−0.6 ± 0.2	+0.1 ± 0.6
	NS	0.05	NS	NS	0.05	0.01	0.02	0.02	NS	NS	0.05	0.05	NS

*Each group of numbers represents the mean value of seven experiments, and consists of mean response ± SEM, mean difference ± SE of difference, and the significance level. NS = nonsignificant. The meaning of the symbols is given in the text.

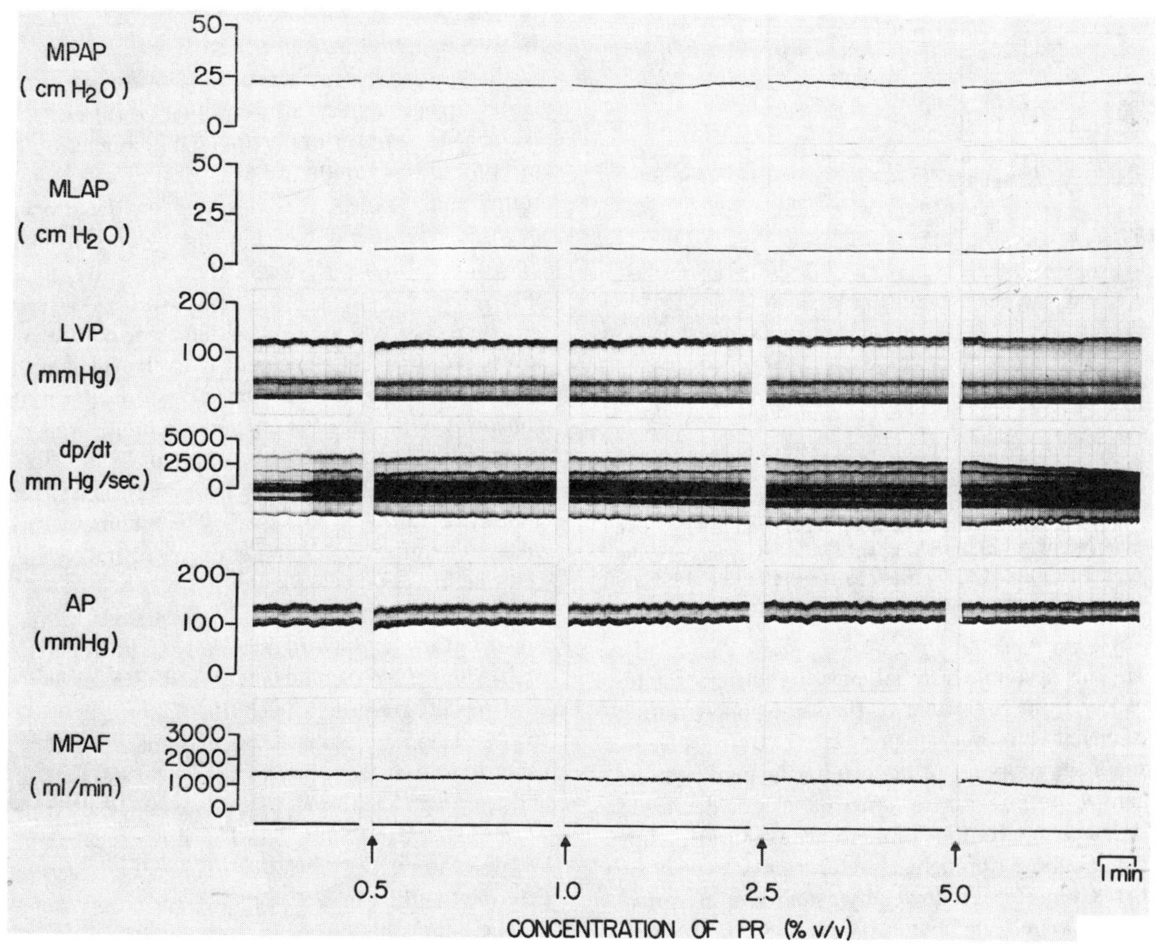

FIGURE 2.8. The effects of progressively increasing concentrations of a paint remover on hemodynamic parameters in a representative experiment. MPAP = mean pulmonary arterial pressure, MLAP = mean left atrial pressure, LVP = left ventricular pressure, dp/dt = maximal rate of rise of left ventricular pressure, AP = aortic pressure, MPAF = mean pulmonary arterial flow.

respectively. A decrease in stroke volume was also induced by inhalation of various concentrations of paint remover, thus a decrease of 2.2%, 3.6%, 4.3%, and 7.2% was induced by 0.5%, 1.0%, 2.5%, and 5.0%, respectively. An initial increase followed by a decrease in heart rate was also elicited by various concentrations of paint remover. No significant changes in other parameters were observed. The parameters that are most affected by inhalation of the paint remover are depicted in Figure 2.9.

Comparison of the paint remover and methylene chloride – Since the paint remover consists of 90.17% w/w of methylene chloride, it was found necessary to compare the various hemodynamic and myocardial changes brought about by the two solvents. The addition of 4.2% methanol, 3.2% isopropanol and 2.4% toluene was expected to modify, i.e., potentiate, synergize, or antagonize the effects of methylene chloride. Results are depicted in Figure 2.10 At lower concentrations, (0.5% and 1.0%) both methylene chloride and paint remover induced the same qualitative effect in so far as the left ventricular and systemic pressures are concerned. Nonetheless, a high concentration (5%) brought about changes in these two parameters that were different in magnitude and in direction. Thus, while methylene chloride brought about a decrease in left ventricular pressure and systemic arterial pressure of 3.9% and 3.7%, respectively, inhalation of the same concentration of paint remover induced an increase in the same parameters of 2.1% and 7.9%, respectively. The effect of 5% methylene chloride

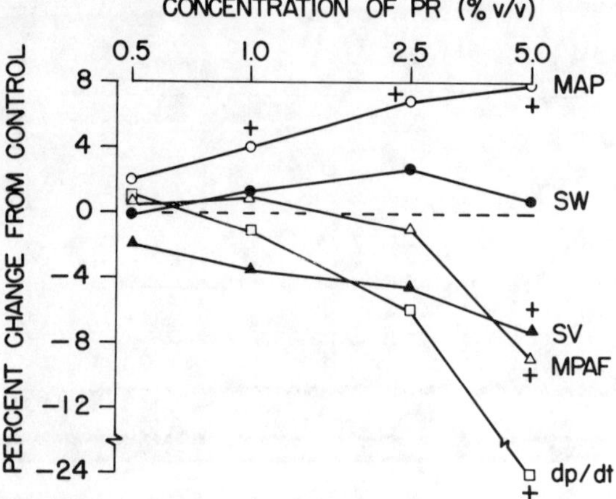

FIGURE 2.9. Concentration-response curve of some of the hemodynamic parameters after the inhalation of paint remover in the anesthetized dog preparation. SV = stroke volume, SW = stroke work, other abbreviations are the same as Figure 1.1. Vertical bars representing SEM are omitted for clarity. + denotes significant change.

on the systemic arterial pressure differs significantly from the effect of the same concentration of paint remover (Figure 2.10). Both solvents share the property of depressing the myocardium; cardiac output, on the other hand, was decreased by all concentrations of methylene chloride, while paint remover brought about a biphasic response; first an increase, followed by a decrease, in cardiac output as indicated by changes in mean pulmonary arterial flow. There is a significant difference between the decrease in cardiac output brought about by methylene chloride and paint remover at 2.5% and 5.0% concentrations. No arrhythmogenic effect was observed with the paint remover.

C. Discussion

1. Acute Toxicity in Mice

Methylene chloride alone – The results of this investigation show that the line correlating the log dose with percentage mortality is steep in the case of methylene chloride. Thus, while 0.65 g/kg oral dose of this solvent killed no mice, doubling the dose gave rise to 22.2% mortality (Table 2.1). This holds true also when other routes of administration were used, viz., intraperitoneally and by inhalation. This is in agreement with observations of other authors who found only a slight difference between the amounts required to induce surgical anesthesia and death.[23] Mortality could be due to the effects of methylene chloride on the central nervous system and/or the cardiovascular system. Since high concentrations of methylene chloride were shown to possess arrhythmogenic properties and sensitized the heart to the arrhythmogenic effect of epinephrine,[24] death was probably due to cardiovascular stress either from methylene chloride itself or its metabolite, carbon monoxide. This conclusion is in contradition to the opinion of Stewart[25] who stated that "except for its action on the central nervous system and the skin, methylene chloride does not exert a significant effect on other organ systems."

The LD_{50} and LC_{50} calculated in the present investigation are in agreement with the findings of other authors. Thus, the oral LD_{50} of 1.99 g/kg conforms with values obtained by Kimura et al.[26] using rats: 14-day-old (1.8 ml/kg or 2.38 g/kg), young adults (1.6 ml/kg or 2.12 g/kg), and older adults (2.3 ml/kg or 3.05 g/kg). The minimal lethal dose 24 hr after oral administration of methylene chloride in dogs was 3 g/kg.[27]

The intraperitoneal LD_{50} calculated in the present study is 448 mg/kg. This is in agreement with findings of Schuhmacher and Grandjean[28] who reported a value of 0.33 to 0.35 ml/kg (407 to 464 mg/kg). Nonetheless, Gradiski et al.[29] using female mice reported a value of 1900 mg/kg, and Gehring[30] reported 23 m*M*/kg (fiducial limits 17 to 31), i. e., 1953 mg/kg. This difference might be due to a sex difference or other factors determining the toxicity of methylene chloride.

The inhalational LC_{50} found in the present investigation is 2.671% v/v or 26,710 ppm for 20 min exposure in mice. This correlates well with findings of other authors, e.g., 14,500 ppm was the LC_{50} after 2 hr exposure,[31,32] and 16,186 ppm after 7 hr exposure.[33] Other values of 17,144 ppm (63 mg/l) and 14,400 ppm (50 mg/l) were reported using static arrangements.[34] Inhalation of 63 mg/l (17,144 ppm) brought about loss of posture in mice after 6 min.[35]

Methylene chloride and paint remover – Comparison of the toxicity of methylene chloride and paint remover showed that by the oral route they possess almost the same LD_{50} (1.987 and 1.997 g/kg, respectively). However, given intraperitoneally or by inhalation, the acute toxicity of the paint remover (PR) is less than that of methylene chloride (MCL), having 59.1% and 74.9% toxicity of the latter. Relative toxicity calculations are based on the formula:

$$\frac{1/LD_{50} \text{ or } 1/LC_{50} \text{ for PR}}{1/LD_{50} \text{ or } 1/LC_{50} \text{ for MCL}} \times 100.$$ The incorporation of methyl alcohol, isopropanol, and

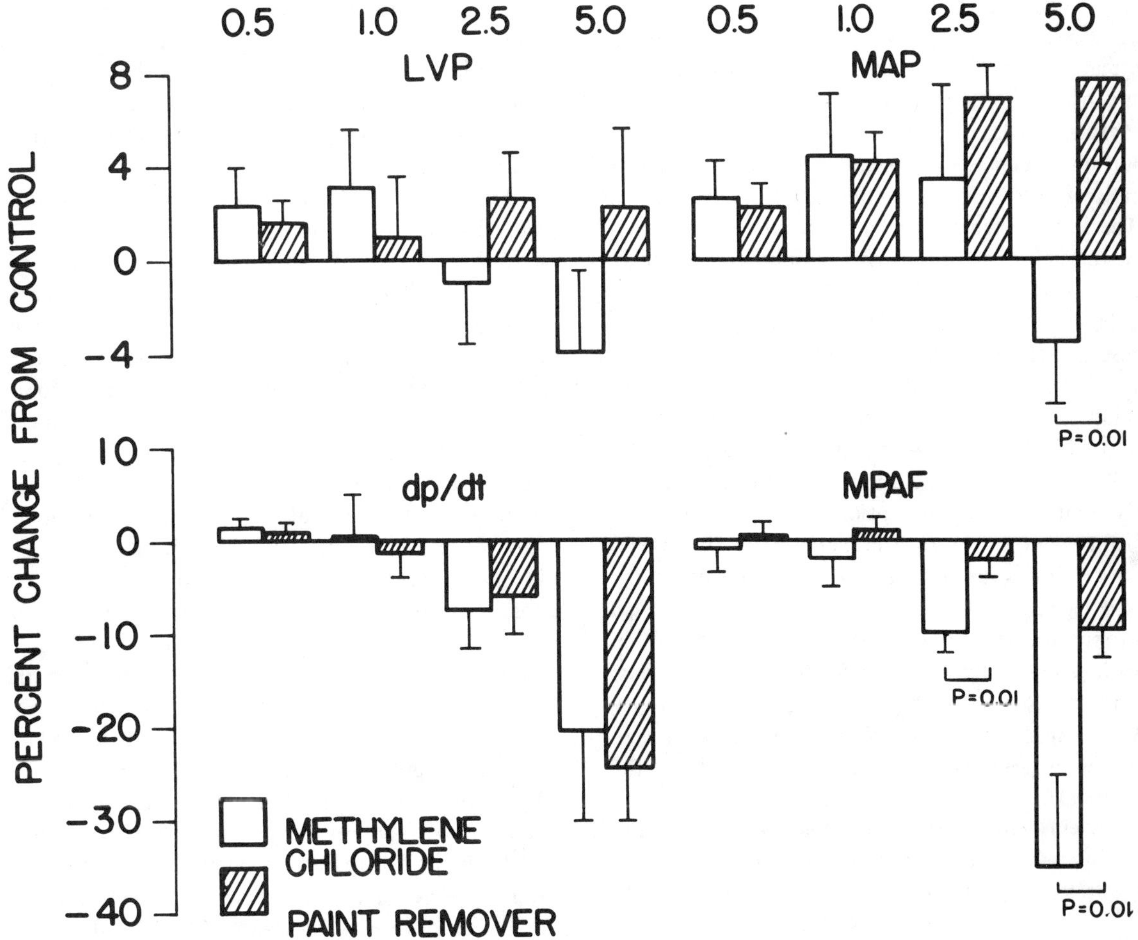

FIGURE 2.10. Comparative effects of methylene chloride and paint remover on left ventricular pressure (LVP), mean aortic pressure (MAP), myocardial contractility (dp/dt), and mean pulmonary arterial flow (MPAF). Concentrations % v/v are indicated at the top.

toluene in the paint remover has, therefore, modified the toxicity of methylene chloride. Alcohols (methanol and isopropanol), being hydroxylated solvents, tend to increase the solubility of the whole mixture in water; in other words, the oil/water partition coefficient and the toxicity of the paint remover is expected to be less than that of methylene chloride. This speculation is based on findings of von Oettingen et al.,[34] who reported on the partition coefficient of methylene chloride and chloroform using water and oleyl alcohol phases. The oil/water distribution coefficient is 25.6 for methylene chloride and 154 for chloroform. The more halogen atoms in the chlorinated methanes, the greater the partition coefficient of these solvents and the more toxic they are. For instance, CCl_4 is more toxic than $CHCl_3$ which, in turn, is more toxic than CH_2Cl_2. The presence of methanol, isopropanol, and toluene in paint remover is expected to change not only the partition coefficient of the whole mixture but also other physicochemical properties such as vapor pressure, boiling point, etc., which might affect the kinetics and eventually the toxicity of the whole mixture. Balmer and associates[36] found that when a combination of methylene chloride and ethanol was inhaled by guinea pigs for 1 day, there was an antagonism between the effects of the two solvents on liver. Hence the possibility exists of an antagonism between the effects of the four components of paint remover, and/or their metabolic products as an explanation for the decrease in toxicity of this mixture.

The oral route of administration, nonetheless, showed no discrimination between the toxicity of methylene chloride and the paint remover. It is well known that lipid-soluble nonelectrolytes, e. g., alcohols, are rapidly absorbed into the blood stream by diffusion across the gastrointestinal mucosae. Consequently, the incorporation of methanol and isopropanol in the paint remover not only decreases its toxicity but also increases its rate of absorption from the gastrointestinal tract. These two effects act in opposition to each other, so the net result will be the algebraic sum of both. It seems that these two contradictory effects are balanced and they annul each other, with a net result of almost no change in toxicity.

The reason for the difference between the relative toxicity of paint remover to that of methylene chloride when given intraperitoneally (59.1%) and by inhalation (74.9%) can only be speculative. The absorption of the mixture of paint remover from the peritoneum is mainly dependent upon its partition coefficient, while when given by inhalation, other physicochemical properties, for instance, the partial pressure of various components, the vapor pressure of the mixture, etc., seem to contribute to the rate of absorption from the lungs. This plausible explanation is in line with results of Lazarew[37] who found that the relative toxicity of methylene chloride (dissolved in water) to that of chloroform is 0.46 (chloroform = 1), while the calculated vapor toxicity of methylene chloride relative to chloroform is 0.79.

2. Hemodynamic Effects of Methylene Chloride in Dogs

Results of these experiments showed that methylene chloride, unlike methyl chloroform, elicited a biphasic response, i. e., increase in left ventricular pressure, left ventricular dp/dt, mean arterial pressure and stroke work, at lower concentrations followed by a decrease in the same parameters at higher concentrations.

Myocardial contractility – That methylene chloride is a myocardial depressant agent is evident from the decrease in the rate of rise of left ventricular pressure, decrease in cardiac output as shown from the decrease in pulmonary arterial flow, increase in mean left atrial pressure, and negative chronotropic effect. The significant decrease in cardiac output appears to be due to the attenuation of myocardial contractile force and, probably, to a decrease in venous return. This is further evidenced by the consistent and concentration-dependent decrease in stroke volume of 4.6, 11.1, and 31.4% following the inhalation of 1.0%, 2.5%, and 5% of methylene chloride, respectively. This myocardial depressant property is common amongst various halogenated and nonhalogenated organic solvents and gases studied in this laboratory and commonly used as solvents and/or propellants in various sprays.

Pulmonary vascular resistance – A nonsignificant decrease in pulmonary vascular resistance was exhibited by 0.5% and 1% concentrations of methylene chloride, and a nonsignifcant increase was induced by 2.5% and 5% of the same compound; this can be explained by examining the effect of methylene chloride on the various components of the pulmonary vascular bed. Since pulmonary vascular resistance is a derived parameter, it can be affected by various factors affecting pulmonary arterial pressure and blood flow. Methylene chloride brings about a decrease in mean pulmonary arterial blood flow at all concentrations studied. This tends to increase the pulmonary vascular resistance. On the other hand, methylene chloride brings about a decrease in effective mean pulmonary arterial pressure at all concentrations studied; this decrease tends to reduce the pulmonary vascular resistance. The net result depends on the relative changes in these two parameters. Thus at concentrations of 0.5% and 1.0% there is a reduction in pulmonary vascular resistance because the decrease in effective mean pulmonary arterial pressure (2.6% and 4.3%, respectively) is greater than the decrease in pulmonary arterial flow (0.2% and 2%, respectively). At higher concentrations of 2.5% and 5% there is an increase in pulmonary vascular resistance because the decrease in pulmonary arterial flow (10.0% and 35.3%) overshadows the reduction in effective mean pulmonary arterial pressure (7.3% and 29.1%). Since effective mean pulmonary arterial pressure is the difference between mean pulmonary arterial pressure and mean left atrial pressure, changes in myocardial contractility will, in turn, influence effective pressures. To add to the confusion, the net result is also modified by possible direct and reflex mediated effects of methylene chloride on the pulmonary vascular bed. The sympathoadrenal discharge as well as the central medullary effect might play a role as well.

Systemic vascular resistance – The systemic vascular resistance is increased by all concentrations of methylene chloride, especially by 2.5% and 5.0%. This could be due largely to the decrease in cardiac output, as changes in mean arterial pressure are nonsignificant. Furthermore, this increase in systemic vascular resistance is the sum total of changes in skeletal muscle, abdominal, renal, cerebral, mucosal, and coronary vascular beds.

Release of catecholamines – The initial increase in left ventricular pressure, left ventricular dp/dt, mean arterial pressure, heart rate, mean pulmonary arterial pressure, and stroke work raised the question that methylene chloride, in lower concentrations, might stimulate the sympathoadrenal system, leading to release of catecholamines which overshadows its myocardial depressant properties that are displayed at higher concentrations. For this reason, a 1% concentration of methylene chloride was administered to the open-chest dog preparation before and after treatment with phentolamine mesylate and propranolol hydrochloride. Figure 2.3 shows that the only parameter that is significantly changed after alpha-adrenergic blockade is the pulmonary arterial pressure. However, after blockade of beta-adrenergic receptors, a significant decrease in left ventricular pressure, left ventricular dp/dt, heart rate, mean arterial pressure, mean pulmonary arterial flow, effective mean pulmonary arterial pressure, stroke volume, and stroke work, was obtained by 1% methylene chloride. This indicates that release of catecholamines from their storage sites is involved in the effect of methylene chloride, which masks its myocardial depressant effects, especially at lower concentrations.

Comparison of methylene chloride and methyl chloroform – The comparative study of the hemodynamic effects of methyl chloroform and methylene chloride showed that, generally, methyl chloroform is effective in lower concentrations; the thresholds for depression of myocardial contractility of these two compounds are 0.1% and 0.5%, respectively. Furthermore, the concentrations that elicited persistent myocardial arrhythmias were 2.5% and 5%, respectively. This is in agreement with previous studies reported by Belej et al.[38] The most outstanding effect of methyl chloroform is on the myocardial contractile force, while the most prominent action of methylene chloride is on the cardiac output. The release of catecholamines from their storage site by methylene chloride with the resultant pooling of blood in the skeletal muscle vascular bed, and the decrease in myocardial contractility might explain this effect.

3. Hemodynamic Effects of Paint Remover in Dogs

In the present investigation, the myocardial and hemodynamic effects of a paint remover (containing 90.17% w/w methylene chloride) were studied in the open-chest dog preparation. As in the case of methylene chloride, the most prominent effect of the paint remover was attenuation of myocardial contractility and decrease in cardiac output.

Myocardial contractility – The initial increase in myocardial contractility and the positive chronotropic effects observed following the inhalation of lower concentrations of paint remover, which are similar to those noticed with methylene chloride alone (Figure 2.10), might be attributed to the stimulation of the sympathoadrenal system which overcomes the decrease in myocardial contractility induced by methylene chloride and other components. The decrease in myocardial contractility observed with higher concentrations of the paint remover is almost identical to and not significantly different from that elicited by methylene chloride alone.

Cardiac output – The effects of methylene chloride alone, and methylene chloride-containing paint remover, on the cardiac output are different in direction (lower concentrations) and in magnitude (higher concentrations). In the absence of detailed myocardial and hemodynamic studies of the other components of the paint remover, viz., methanol, isopropanol, and toluene, the explanation of these differences can only be speculative. The net effect of the paint remover on the cardiac output is the algebraic sum of several factors that are different in magnitude and in direction. Among these factors is the direct effect of various components of paint remover and/or their metabolites on the cerebral, renal, mesenteric, pulmonary, coronary, skeletal muscle, and cutaneous vascular beds, and possibly on the heart. The various hemodynamic compensatory mechanisms that are not directly evoked by the solvents may modify or even annul the direct hemodynamic effects. The possible vagal influences activated reflexly by the local irritant effect of these components on the tracheobronchial tree cannot be ignored. The picture is further complicated by the central depressant effect of methylene chloride and/or other solvents present in the paint remover that tend to inhibit the vasomotor center with consequent vasodilation. The attenuation of myocardial contractility tends to decrease systolic ventricular pressure and cardiac output and increase ventricular end-diastolic pressure. On the other hand, an increase in the venous tone tends to augment the venous return which, in turn, increases the ventricular systolic pressure and the ventricular end-diastolic pressure. The net result of these two antagonistic forces would, therefore, be an increase in the ventricular end-diastolic pressure (since both forces act in the same direction) and a variable response on the ventricular pressure (depending upon the algebraic sum of these contradictory effects). This speculation is not difficult to appreciate in the light of the present results which showed a gradual rise in left ventricular end-diastolic pressure and almost no change in left ventricular systolic pressure (Table 2.9).

Systemic arterial pressure – The effect of the paint remover on the systemic arterial pressure deserves a comment. A decrease in cardiac inotropism and output induced by inhalation of paint remover tends to decrease the mean arterial pressure. However, a consistent and significant increase in arterial pressure was observed with all concentrations studied (see Figure 2.9 and Table 2.9). This effect implies that the net result of various components of the paint remover and/or its metabolites is an increase in the vascular tone of the arterioles. This effect tends to increase the vascular resistance which not only nullifies the decrease in cardiac output but even brings about a net increase in systemic pressure. This effect is indirectly aided by the increase in venous tone which increases venous return and, coupled with the decrease in myocardial contractility, would ultimately lead to a relatively small decrease in cardiac output.

Cardiac arrhythmia – Another feature in the pharmacologic spectrum of the paint remover is the absence of myocardial arrhythmia even though methylene chloride is contained in the paint remover. The arrhythmogenic effect of methylene chloride was demonstrated in our experiments on anesthetized dogs (Figure 2.7) and by other investigators using nonanesthetized dog.[39,40] The role of the constituents of the paint remover (methanol, isopropanol, toluene, or one or more of their metabolic products) in reducing the arrhythmogenic effect of methylene chloride needs further investigation.

D. Summary

The inhalation of 0.5 to 5.0% of methylene chloride in the anesthetized dog results in the following: (a) initial increase in left ventricular pressure, left ventricular dp/dt, mean arterial pressure, and stroke work caused by release of catecholamines; (b) decrease of the same parameters due to direct depression of myocardial contractility; (c) decrease in cardiac output brought about by myocardial depression and probably decrease in venous return; (d) increase in pulmonary vascular resistance as a consequence of the reduced output; (e) increase in systemic vascular resistance in part due to release of catecholamines; and (f) cardiac arrhythmia. In contrast to methylene chloride, lower concentrations of methyl chloroform (0.1 to 1.0%) elicit the same cardiac effects, (b), (c), and (e). Methyl chloroform does not cause initial stimulation of cardiac function, causes a more conspicuous increase in pulmonary vascular resistance, and decreases systemic vascular resistance.

Both methylene chloride and a paint remover (containing 90.17% w/w methylene chloride, 4.2% w/w methanol, 3.24% w/w isopropanol, and 2.4% w/w toluene) decreased the myocardial contractility and output in the anesthetized open-chest dog preparation. However, some quantitative and qualitative differences in the pharmacologic and toxicologic spectrum of these two agents have been revealed in the above-mentioned preparation as well as in mice. These are as follows: (a) while methylene chloride brought about a biphasic response, viz., increase followed by decrease in left ventricular pressure and systemic arterial pressure, the paint remover produced an increase in these parameters in concentrations up to 5%; (b) the paint remover brought about an increase in systemic vascular resistance more than that elicited by methylene chloride; (c) the paint remover is not arrhythmogenic, in contrast to methylene chloride; (d) administered intraperitoneally or by inhalation to mice, the paint remover is less toxic than methylene chloride; orally, they are almost the same insofar as acute toxicity is concerned.

Chapter 3

ETHANOL

The use and abuse of ethanol in the form of alcoholic beverages has stirred the interest of various investigators in studying the pharmacodynamic and toxic effects of ethanol on the central nervous and cardiovascular systems, as well as on the metabolism of various animals and man.[41,42] In these studies, ethanol was administered orally, infused intravenously, or added to the isolated organ bath, but not inhaled.

Ethanol is extensively used as a solvent in aerosol products. Among its numerous uses in medicine, ethanol is administered by inhalation as an antifoaming agent in the management of acute pulmonary edema secondary to left heart failure.[43] Browning[44] stated that "...inhalation of the vapour of ethyl alcohol to the point of intoxication produces the same effects as a toxic dose taken by the mouth." The toxic effects of inhaled ethanol on the central nervous system were studied in rats and guinea pigs.[45] In experiments in man, Browning[44] observed that the inhalation of 1.9 mg/l of ethanol produced slight symptoms of poisoning; 9.5 mg/l caused strong stupor and morbid drowsiness. However, the hemodynamic effects of ethanol administered by inhalation have not hitherto been reported on, to the best of our knowledge.

Since myocardial depression is a well-known feature of acute ethanol intoxication,[46-49] and in light of the above-mentioned facts, it was found necessary to study the effect of ethanol administered by inhalation on the cardiovascular system of the anesthetized dog preparation.

A. Methods and Calculations

The experiments were performed on adult mongrel dogs of either sex, weighing 18 to 24 kg (average 20.8) which were anesthetized by intravenous administration of 30 to 35 mg/kg pentobarbital sodium. The details of measurements of hemodynamic parameters, calculations, and abbreviations used are described in Chapter 2.

About 30 min elapsed before the administration of ethanol vapors, during which the preparation was allowed to stabilize. The following concentrations of ethanol in air were then administered in succession, via the inlet of the respirator, each for 5 min: 1.0% (18.84 mg/l), 2.5% (47.11 mg/l), 5.0% (94.21 mg/l), 7.5% (141.32 mg/l), and 10% (188.42 mg/l).

Various concentrations of ethanol were prepared by volatilizing the appropriate amount of the solvent, calculated to give the appropriate volume of vapors at standard pressure and 25°C, as previously described.[73]

Results are represented as mean ± SEM. Student's t-test modified for paired replicates was used to compare data obtained from the same animal. In all cases, P values of less than 0.05 were considered significant.

B. Results

The response of the open-chest dog preparation to inhalation of various concentrations of ethanol in a typical experiment is illustrated in Figure 3.1, and results are summarized in Table 3.1. Four stages appeared consecutively:

Reduced mean pulmonary arterial flow – A reduction in cardiac output (5.1%) accompanied by reduction in stroke volume (4.2%), stroke work (4.3%) and increase in pulmonary and systemic vascular resistances (7.7 and 4.7%, respectively) were observed. These effects are apparent during the inhalation of 1% ethanol and occur without any detectable change in heart rate or contractility. The hemodynamic change appears to be brought about by an extracardiac mechanism, probably a reduction in venous return.

Pulmonary vasoconstriction – Pulmonary vasoconstriction is reflected by further increase in pulmonary vascular resistance (12.3%), and a rise in mean pulmonary arterial pressure (3.6%) and effective mean pulmonary arterial pressure (5.5%). This state appears during the inhalation of 2.5% ethanol and is explained by a local action on the pulmonary blood vessels. Until this stage, the increase in pulmonary vascular resistance seen during the stage of reduced blood flow is an indirect outcome of the reduced pulmonary blood flow.

Myocardial depression – With inhalation of 7.5% ethanol, significant depression of contractility (9.6%) appears for the first time accompanied by an increase in heart rate (1.9%). Myocardial depression was further intensified by increase in inhaled concentration of ethanol. It is noteworthy that inhalation of 1% ethanol elicited an increase that was statistically nonsignficant in the left ventricular dp/dt. A concentration-dependent nonsignificant increase in left ven-

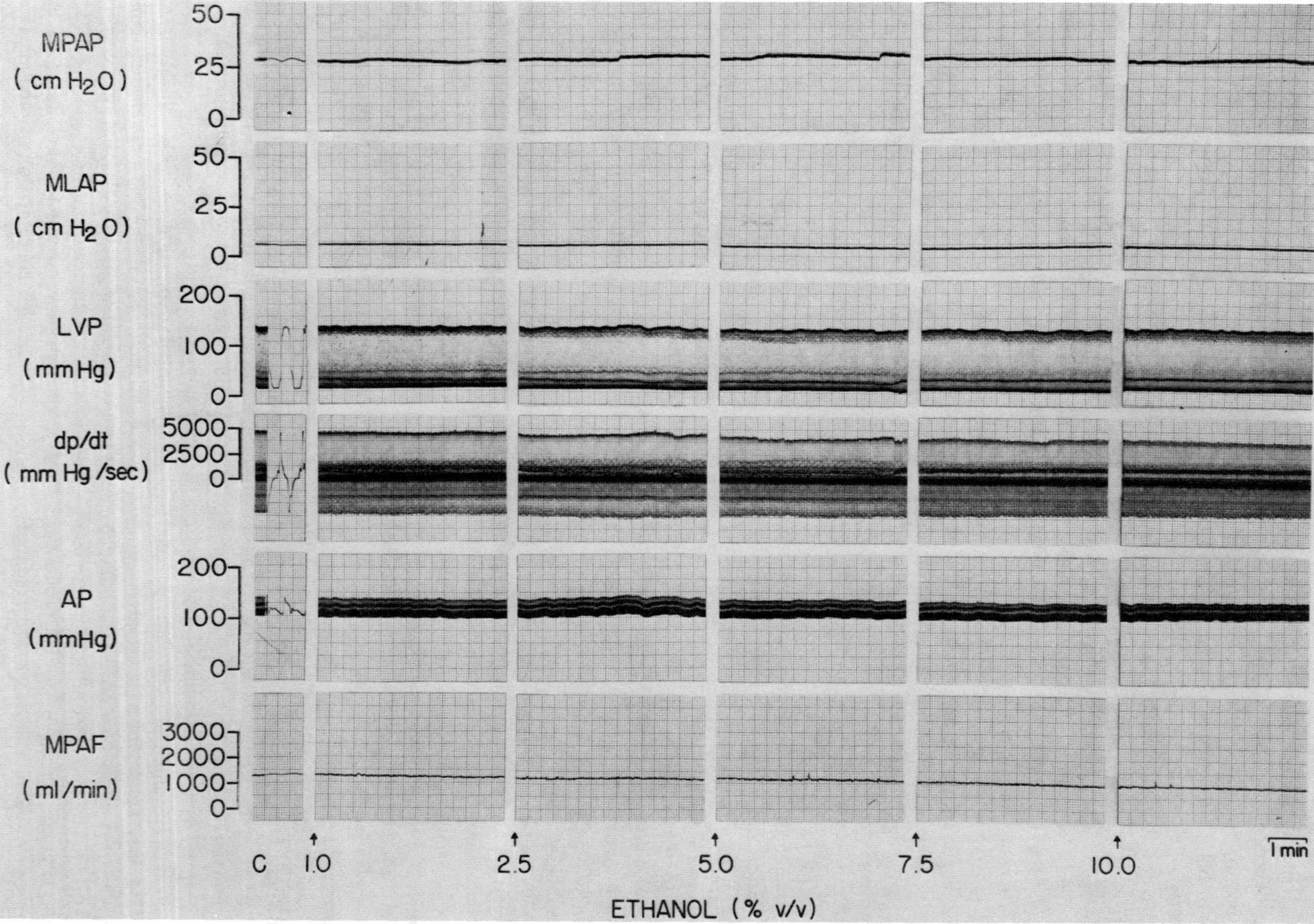

FIGURE 3.1. The effects of progressively increasing concentrations of ethanol on the hemodynamics of the open-chest dog preparation. MPAP = mean pulmonary arterial pressure; MLAP = mean left atrial pressure; LVP = left ventricular pressure; dp/dt = maximal rate of rise of left ventricular pressure; AP = aortic pressure; MPAF = mean pulmonary arterial flow.

TABLE 3.1

Hemodynamic Effects of Inhalation of Progressively Increasing Concentrations of Ethanol in Anesthetized Dogs*

	MEAP cm H_2O	MLAP cm H_2O	EMPAP cm H_2O	LVP mmHg	LVEDP mmHg	dp/dt mmHg/sec	MAP mmHg	MPAF ml/min	HR beats/min	Vascular resistance dynes·sec/cm^5		Stroke vol ml	Stroke work g·meter
										pulmonary	systemic		
Control	33.8 ± 2.3	7.5 ± 1.1	25.4 ± 2.3	141 ± 8.3	2.5 ± 1.4	4313 ± 410	125 ± 9.1	1650 ± 181	174 ± 7.3	972 ± 122	6036 ± 397	9.6 ± 1.1	17.4 ± 3.0
1%	34.3 ± 2.7 +0.5 ± 0.7 NS	7.4 ± 1.3 −0.2 ± 0.2 NS	27.0 ± 2.5 +0.6 ± 0.6 NS	144 ± 10.0 +2.3 ± 2.8 NS	3.3 ± 2.0 +0.8 ± 0.8 NS	4406 ± 403 +94 ± 31 NS	125 ± 9.8 0 ± 1.2 NS	1563 ± 159 −88 ± 23.9 0.05	173 ± 8.3 −1.0 ± 1.2 NS	1048 ± 133 +76 ± 22.4 0.05	6306 ± 337 +270 ± 72.9 0.05	9.1 ± 1.0 −0.4 ± 0.1 0.05	16.6 ± 2.9 −0.8 ± 0.2 0.05
2.5%	35.0 ± 2.4 +1.2 ± 0.2 0.01	7.2 ± 1.3 −0.3 ± 0.2 NS	27.8 ± 2.3 +1.4 ± 0.1 0.01	147 ± 11.3 +5.3 ± 4.5 NS	4.5 ± 3.1 +2.0 ± 2.0 NS	4313 ± 373 0 ± 51 NS	125 ± 10.5 0 ± 1.8 NS	1550 ± 167 −100 ± 20.4 0.02	174 ± 6.8 +0.5 ± 1.0 NS	1090 ± 132 +118 ± 13.8 0.01	6322 ± 414 +286 ± 99.9 NS	9.0 ± 1.0 −0.6 ± 0.1 0.02	16.3 ± 3.0 −1.1 ± 0.2 0.01
5.0%	36.1 ± 2.3 +2.3 ± 0.3 0.01	7.4 ± 1.3 −0.2 ± 0.2 NS	28.7 ± 2.1 +2.3 ± 0.4 0.01	143 ± 11.8 +1.3 ± 4.7 NS	6.3 ± 3.2 +3.8 ± 2.4 NS	4000 ± 335 −313 ± 120 NS	123 ± 11.1 −2.5 ± 2.1 NS	1500 ± 177 −150 ± 28.9 0.02	176 ± 6.8 +2.5 ± 1.3 NS	1170 ± 140 +197 ± 45.9 0.02	6352 ± 535 +316 ± 231.9 NS	8.6 ± 1.0 −1.0 ± 0.1 0.01	15.3 ± 2.9 −2.1 ± 0.2 0.01
7.5%	35.8 ± 3.0 +2.0 ± 0.9 NS	7.4 ± 1.3 −0.2 ± 0.2 NS	28.4 ± 2.6 +2.0 ± 0.8 NS	140 ± 9.8 −1.8 ± 2.4 NS	6.3 ± 3.2 +3.8 ± 2.4 NS	3906 ± 393 −406 ± 60 0.01	123 ± 11.1 −2.5 ± 2.1 NS	1425 ± 185 −225 ± 32.3 0.01	177 ± 6.8 +3.3 ± 1.0 0.05	1219 ± 149 +246 ± 33.7 0.01	6734 ± 636 +699 ± 351.1 NS	8.1 ± 1.0 −1.5 ± 0.2 0.01	14.5 ± 2.9 −2.9 ± 0.4 0.01
10.0%	37.2 ± 3.2 +3.4 ± 1.1 NS	7.4 ± 1.3 −0.2 ± 0.2 NS	29.8 ± 2.6 +3.4 ± 1.0 0.05	140 ± 9.6 −1.3 ± 2.4 NS	5.8 ± 1.5 +3.3 ± 1.2 NS	3656 ± 312 −656 ± 107 0.01	122 ± 11.5 −3.3 ± 2.7 NS	1375 ± 185 −275 ± 32.3 0.01	180 ± 6.7 +6.0 ± 1.5 0.05	1325 ± 151 +352 ± 35.6 0.01	6963 ± 717 +928 ± 421.2 NS	7.7 ± 1.0 −1.9 ± 0.2 0.01	13.8 ± 2.8 −3.6 ± 0.4 0.01

*Each group of numbers includes mean ± standard error for five dogs, mean difference ± standard error, and the significance level.

tricular and diastolic pressure was observed with all concentrations of inhaled ethanol.

Systemic hypotension – Levels of ethanol up to 10% did not reduce aortic blood pressure. A concentration of 15% was required to induce hypotension. The results are based on two dogs and have been omitted from Table 3.1. Nonetheless, a significant increase in heart rate was observed with concentrations of 7.5% and 10% of ethanol.

C. Discussion

The experiments reported above are the first hemodynamic investigation of ethanol administered by inhalation in the anesthetized open-chest dog preparation. The cardiovascular effects of various doses of ethanol given by intravenous infusion in the open-chest, closed-chest, conscious, or anesthetized dog preparation are summarized in Table 3.2.

Effect on cardiac output – Table 3.2 reveals much contradiction between results obtained by various authors, especially insofar as the cardiac output is concerned. These contradictory results can be explained on the grounds of variation in experimental conditions, alcohol concentration and/or dose infused, and in the interval of observation of hemodynamic changes following the infusion. In the present investigation, however, there is a consistent and significant decrease in the cardiac output, elicited by as low a concentration as 1%. Since at this concentration there is no concomitant significant decrease in heart rate nor in myocardial contractility, as reflected from the left ventricular dp/dt, it seems most probable that this decrease in cardiac output is due to decrease in ventricular filling pressure which in turn is due to decrease in venous return.

It is evident from Table 3.1 that there is an increase in left ventricular end diastolic pressure; though statistically nonsignificant, it reflects the decrease in myocardial contractility that overcomes the decrease in venous return. Since ethanol has an insignificant direct effect on the blood vessels[50] the central vasomotor depression brought about by ethanol, with the consequent pooling of blood in the periphery that ultimately leads to decrease in venous return, might play the major role in decreasing the cardiac output. At higher concentrations, however, this decrease in cardiac output is further intensified by the depression of myocardial contractility and decrease in systolic ejection.

Pulmonary vasoconstriction – The use of alcohol by inhalation in acute pulmonary edema depends on its surface action that collapses the foam in the tracheobronchial airway. Along similar lines, ethanol, by virtue of its surface active properties, might change the permeability of the pulmonary capillaries, leading to capillary congestion and rise in pulmonary arterial pressure, similar to that produced by alloxan.[51] The rise in pulmonary arterial pressure, though nonsignificant, is concentration-dependent and consistent. A direct effect on lung parenchyma, J-receptors, or the release of catecholamines might be involved. The latter mechanism, viz., release of catecholamines from their storage sites, is not difficult to appreciate knowing that ethanol is metabolized by alcohol dehydrogenase to acetaldehyde, which was found to release catecholamines from the adrenal medulla.[52,53]

Depression of myocardial contractility – Ethanol-induced depression of myocardial contractility as directly measured by the strain gauge,[54] or as reflected from the maximum rate of rise of left ventricular dp/dt,[55-62] constitutes the most consistent response observed by various authors (see Table 3.2). This holds true in the case of conscious animals,[57,59] or in the case of anesthetized dogs in which a certain level of myocardial depression is achieved by pentobarbital. Furthermore, myocardial depression is demonstrated in vitro using the isolated heart preparation.[46,48,63,64] In patients with coronary heart disease, a decrease in myocardial contractility was also reported.[42] On the other hand, stimulation of the mammalian heart was reported in 1905 by Loeb,[65] Bachem,[66] and in 1907 by Dixon.[67]

In the present investigation, myocardial contractility was gauged by the changes in the maximum rate at which ventricular pressure is developed (dp/dt). This method gives a sensitive and valid measure of the contractile state of the ventricle in the intact dog preparation so far as the preload and after load are controlled.[68] Changes in left ventricular end-diastolic pressure and in aortic pressure, in the present experiments were nonsignificant and would, therefore, not account for the large and significant changes in the dp/dt, which, in turn, reflects myocardial depression brought about by ethanol. This direct myocardial depressant action may account for the anti-arrhythmic effect of ethanol reported by Kostis et al.[69] and the suppression of ventricular tachycardia due to coronary ligation or acetylstrophan-

TABLE 3.2

Summary of the Hemodynamic Effects of Various Doses of Ethanol in the Open- and Closed-chest Dog Preparation

Ethanol	HR	MAP	LVP	LVEDP	SVR	MFC	LVdp/dt	SER	CO	SV	SW	Cor.F.	Cor.R.	Reference
70 mg %											↑	↓	↑	56
70-120 mg %									↑			→		
65-150 mg %		↓										↑	↓	
273 mg %		→				↓								54
1 g/kg*	-	↓			↓				↑			↑	↓	59
120 mg %*	-			↑			↓			↓				57
311 mg %	-		↑	↑			↓			↓				
43 mg %			↓				↓							
67 mg %	-	↓			↓			↑	↑	↑		↓	↓	58
190 mg %	-	↓			↓			↑	↑	↑		↑	↓	
195 mg %			↓				↓							
58 mg/kg	-	→			↑				↓	↓	↓	↑	↓	55
500 mg/kg		↑			↓				↑	↑	↑	↓	↑	61
1500 mg/kg		↑			↓				↑	↑	↑	↓	↑	
5000 mg/kg		↓							↓			↓		
100 mg %	→	↓		↑					↓					
200 mg %	-	↑		↑					↓					62
334 mg %	→			↑			↓		↓	↓				60
200 mg %	-	↑				↑								52, 72
300 mg %	-	↓				↓								

*= Conscious dogs; the headings are abbreviated as follows: HR = heart rate, MAP = mean aortic pressure, LVP = left ventricular pressure, LVEDP = left ventricular end diastolic pressure, SVR = systemic vascular resistance, MFC = myocardial force of contraction, LVdp/dt = maximal rate of rise of left ventricular pressure, SER = systolic ejection rate, CO = cardiac output, SV = stroke volume, SW = stroke work, Cor. F. = coronary flow, and Cor. R. = coronary resistance.

thidin observed by others.[70,71] Its weak local anesthetic effect and its effect on sodium conductance[69] may give an alternate explanation for the anti-arrhythmic action of alcohol. The effect of ethanol on myocardial contractility, as reflected by changes in left ventricular dp/dt, and on the heart function as reflected by changes in stroke work versus changes in left ventricular end-diastolic pressure is depicted in Figure 3.2.

Effect on systemic blood pressure and heart rate – Ethanol, administered by inhalation in various concentrations up to 10%, failed to produce any significant decrease in systemic blood pressure. Nonetheless, 7.5% and 10% concentrations produced a decrease of 2.5 and 3.3 mmHg in mean arterial pressure, respectively. Interestingly enough, there was no significant decrease in mean arterial pressure, despite the significant decrease in cardiac output. This might be explained by an increase in total peripheral resistance brought about by ethanol or by its metabolites.

On the other hand, the significant increase in heart rate elicited by the same concentrations may be due either to reflexes from baroreceptors or to release of catecholamines.

Differences between ethanol and isopropanol – The hemodynamic effects of the inhalation of isopropanol have been reported in an earlier volume of this series.[2] Since the same techniques were applied to the present investigation of ethanol, it is now possible to compare the two solvents. The differences in hemodynamic effects are as follows: (1) isopropanol depresses myocardial contractility at a lower concentration than that of ethanol (1.0% and 7.5%, respectively); (2) the concentration of isopropanol which reduces systemic arterial blood pressure is lower than that of ethanol (7.5% and 15%, respectively); (3) to the contrary, the concentration of isopropanol which reduces cardiac output is higher than that of ethanol (2.5% and 1.0%). In the case of ethanol, it seems most probable that the decrease in cardiac output might be due mainly to central vasomotor depression, which brings about pooling of blood in the periphery with a consequent decrease in venous return. In the case of isopropanol, the decrease in cardiac output might be due mainly to depression of myocardial contractility and decrease in systolic ejection. This goes hand in hand with statement (2), in which ethanol was said to have produced a decrease in systemic arterial pressure at a higher concentration than isopropanol; and (4) ethanol has a primary influence on venous return and on pulmonary blood vessels which is lacking with isopropanol. These differences are being uncovered for the first time and suggest the necessity for additional investigation to understand why isopropanol and ethanol have qualitative as well as quantitative differences.

D. Summary

The inhalation of ethanol (1 to 10%) elicited the following myocardial and hemodynamic changes in the anesthetized open-chest dog: (a) decrease in cardiac output; (b) increase in pulmonary arterial pressure and pulmonary vascular resistance; (c) decrease in myocardial contractility; and (d) tachycardia.

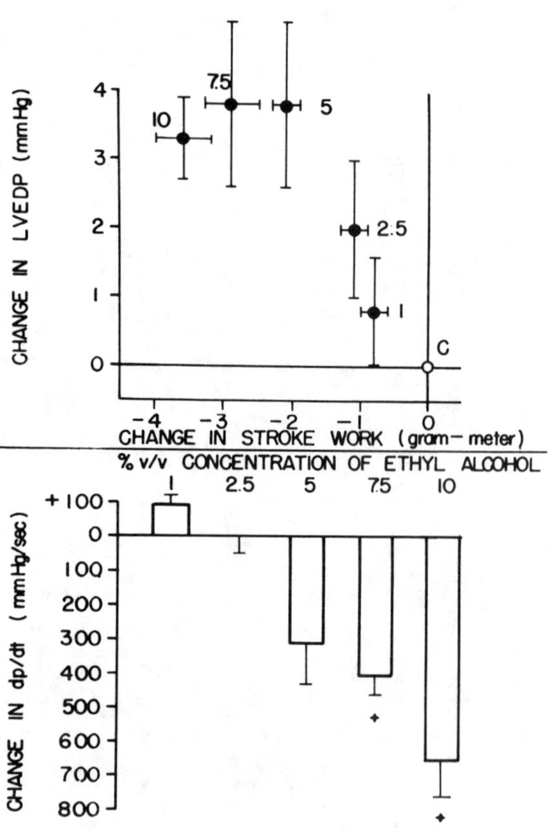

FIGURE 3.2. The effects of progressively increasing concentrations of ethanol on myocardial function (above) and contractility (below). A significant change (+) in myocardial contractility was brought about by 7.5% and 10% concentrations of ethanol.

REFERENCES – Solvents for Aerosols

1. **Aviado, D. M., Zakhari, S., Simaan, J. A., and Ulsamer, A. G.,** *Methyl Chloroform and Trichloroethylene in the Environment,* CRC Press, Cleveland, 1976.
2. **Zakhari, S., Leibowitz, M., Levy, P., and Aviado, D. M.,** *Isopropanol and Ketones in the Environment,* CRC Press, Cleveland, 1976.
3. **Watanabe, T., Ratcliffe, H., Thompson, G., and Aviado, D. M.,** *Fluorinated Propellants for Aerosols,* CRC Press, Cleveland, in press, 1977.
4. **Aviado, D. M.,** Toxicity of propellants, in *Progress in Drug Research,* Jucker, E., Ed., Birkhauser Verlag, Basel, 1974, 365.
5. **Aviado, D. M. and Watanabe, T.,** Functional and biochemical effects on the lungs following inhalation of cigarette smoke and constituents. I. High- and low-nicotine cigarettes in mice, *Toxicol. Appl. Pharmacol.,* 30, 185, 1974.
6. **Folch, J., Lees, M., and Stanley, F. H. S.,** A simple method for the isolation and purification of total lipids from animal tissue, *J. Biol. Chem.,* 226, 497, 1957.
7. **Seifter, J., Dayton, J., Novie, B., and Muntwyler, E.,** The estimation of glycogen with the anthone reagent, *Arch. Biochem.,* 25, 191, 1950.
8. **Horiguchi, S. and Horiuchi, K.,** Report on the toxicity of 1,1,1-trichloroethane, *Jpn. J. Ind. Health,* 13, 226, 1971.
9. **McNutt, N. S., Amster, R. L., McConnell, E. E., and Morris, F.,** Hepatic lesions in mice after continuous inhalation exposure to 1,1,1-trichloroethane, *Lab. Invest.,* 32, 642, 1975.
10. **Adams, E. M., Spencer, H. C., Rowe, V. K., and Irish, D. D.,** Vapor toxicity of 1,1,1-trichloroethane (methyl chloroform) determined by experiments on laboratory animals, *Arch. Ind. Hyg. Occup. Med.,* 1, 225, 1950.
11. **Stewart, R. D., Gay, H. H., Erley, D. S., Hake, C. L., and Schaffer, A. W.,** Human exposure to 1,1,1-trichloroethane vapor – Relationship of expired air and blood concentrations to exposure and toxicity, *Am. Ind. Hyg. Assoc. J.,* 22, 252, 1961.
12. **Torkelson, T. R., Oyen, F., McCollister, D. D., and Rowe, V. K.,** Toxicity of 1,1,1-trichloroethane as determined on laboratory animals and human subjects, *Am. Ind. Hyg. Assoc. J.,* 19, 353, 1958.
13. **Prendergast, J. A., Jones, R. A., Jenkins, L. J., Jr., and Siegel, J.,** Effects on experimental animals of long-term inhalation of trichloroethylene, carbon tetrachloride, 1,1,1-trichloroethane, dichlorodifluoromethane, and 1,1-dichloroethylene, *Toxicol. Appl. Pharmacol.,* 10, 270, 1960.
14. **MacEwen, J. D., Kinkead, E. R., and Haun, C. C.,** A study of the biological effect of continuous inhalation exposure of 1,1,1-trichloroethane (Methyl chloroform) on animals, *NASA,* CR 134323, 1974.
15. **Tsapko, V. G. and Rappoport, M. B.,** The action of methyl chloroform vapors on animals, *Farmakol. Toksikol.* (Moscow), 7, 149, 1972.
16. **Lal, H. and Shah, H. C.,** Effect of methyl chloroform inhalation on barbituate hypnosis and hepatic drug metabolism in male mice, *Toxicol. Appl. Pharmacol.,* 17, 625, 1970.
17. **Fuller, G. C., Olshan, A., Puri, S. K., and Lal, H.,** Induction of hepatic drug metabolism in rats by methyl chloroform inhalation, *J. Pharmacol. Exp. Ther.,* 175, 311, 1970.
18. **Carlson, G. P.,** Effect of phenobarbital and 3-methyl cholenthrene pretreatment on the hepatotoxicity of 1,1,1-trichloroethane and 1,1,2-trichloroethane, *Life Sci.,* 13, 67, 1973.
19. **Stewart, R. D., Gay, H. H., Schaffer, A. W., Erley, D. S., and Rowe, V. K.,** Experimental human exposure to methyl chloroform vapor, *Arch. Environ. Health,* 19, 467, 1969.
20. **Dornette, W. H. L. and Jones, J. P.,** Clinical experience with 1,1,1-trichloroethane – A preliminary report of 50 anesthetic administrations, *Anesth. Analg.* (Cleveland), 39, 259, 1960.
21. **Seki, Y., Urashima, Y., Aikawa, H., Matsumura, H., Ichikawa, Y., Hiratsuka, F., Yoshioka, Y., Shimbo, S., and Ikeda, M.,** Trichloro-compounds in the urine of humans exposed to methyl chloroform at subthreshold levels, *Int. Arch. Arbeitsmed.,* 35, 39, 1975.
22. **Weitbrecht, U.,** Tri and a tri-substitute in the metal industry, *Zentralbl. Arbeitsmed. Arbeitsshutz.,* 15, 138, 1965.
23. **Moskowitz, S. and Shapiro, H.,** Fatal exposure to methylene chloride vapor, *Arch. Ind. Hyg. Occup. Med.,* 6, 116, 1952.
24. **Aviado, D. M. and Belej, M. A.,** Toxicity of aerosol propellants on the respiratory and circulatory systems. I. Cardiac arrhythmia in the mouse, *Toxicology,* 2, 31, 1974.
25. **Stewart, R. D.,** Poisoning from chlorinated hydrocarbon solvents, *Am. J. Nurs.,* 67, 85, 1967.
26. **Kimura, E. T., Ebert, D. M., and Dodge, P. W.,** Acute toxicity and limits of solvent residue for sixteen organic solvents, *Toxicol. Appl. Pharmacol.,* 19, 699, 1971.
27. **Barsoum, G. S. and Saad, K.,** Relative toxicity of certain chlorine derivatives of the aliphatic series, *Q. J. Pharm. Pharmacol.,* 7, 205, 1934.
28. **Schuhmacher, H. and Grandjean, E.,** Comparative investigations into the narcotic activity and the acute toxicity of nine solvents, *Arch. Gewerbepathol. Gewerbehyg.,* 18, 109, 1960.
29. **Gradiski, D., Magadur, J.-L., Baillot, M., Danière, M. C., and Schuh, M. B.,** Toxicité comparée des principaux solvants chlorés aliphatiques, *Eur. J. Toxicol.,* 7, 247, 1974.
30. **Gehring, P. J.,** Hepatotoxic potency of various chlorinated hydrocarbon vapors relative to their narcotic and lethal potencies in mice, *Toxicol. Appl. Pharmacol.,* 13, 287, 1968.

31. **Flury, F. and Zernik, F.,** *Schädliche Gase,* Springer-Verlag, Berlin, 1931, 311.
32. **MacEwen, J. D. and Vernot, E. H.,** Toxic hazards research unit annual technical report, *U.S.N.T.I.S. AD. Rep.,* Issue No. 734543, 69, 1971.
33. **Svirbely, J. L., Highman, B., Alford, W. C., and Von Oettingen, W. F.,** The toxicity and narcotic action of mono-chloro-mono-bromo-methane with special reference to inorganic and volatile bromide in blood, urine and brain, *J. Ind. Hyg. Toxicol.,* 29, 382, 1947.
34. **Von Oettingen, W. F., Powell, C. C., Sharpless, N. E., Alford, W. C., and Pecora, L. J.,** Relation between the toxic action of chlorinated methanes and their chemical and physicochemical properties, *Nat. Inst. Health Bull.,* 191, 1949.
35. **Müller, J.,** Vergleichende Untersuchungen über die narkotische und toxische Wirkung einiger Halogen-Kohlenwasserstoffe, *Arch. Exp. Pathol. Pharmakol.,* 109, 276, 1925.
36. **Balmer, M. F., Smith, F. A., Leach, L. J., and Yile, C. L.,** Inhalation by guinea pigs of methylene chloride alone and in combination with ethyl alcohol, *Am. Ind. Hyg. Assoc. J.,* in press, 1976.
37. **Lazarew, N. W.,** Uber die narkotische Wirkungsdraft der Dampfe der Chlorderivaten des Methans, des Athans und des Athylens, *Naunyn-Schmiedeberg's Arch. Exp. Pathol. Pharmakol.,* 141, 19, 1929.
38. **Belej, M. A., Smith, D. G., and Aviado, D. M.,** Toxicity of aerosol propellants in the respiratory and circulatory systems. IV. Cardiotoxicity in the monkey, *Toxicology,* 2, 381, 1974.
39. **Reinhardt, C. F., Azar, A., Maxfield, M. E., Smith, P. E., Jr., and Mullin, L. S.,** Cardiac arrhythmias and aerosol "sniffing," *Arch. Environ. Health,* 22, 265, 1971.
40. **Reinhardt, C. F., Mullin, L. S., and Maxfield, M. E.,** Epinephrine-induced cardiac arrhythmia potential of some common industrial solvents, *J. Occup. Med.,* 15, 953, 1973.
41. **Tremolieres, J., Ed.,** Alcohol and derivatives, Sec. 20, *International Encyclopedia of Pharmacology and Therapeutics,* Vol. 1 and 2, Pergamon Press, Oxford, 1970.
42. **Regan, T. J.,** Ethyl alcohol and the heart, *Circulation,* 44, 957, 1971.
43. **Luisada, A. A., Goldman, M. A., and Weyl, R.,** Alcohol vapor by inhalation in the treatment of acute pulmonary edema, *Circulation,* 5, 363, 1952.
44. **Browning, E.,** *Toxicity of industrial organic solvents,* Her Majesty's Stationery Office, London, 1953, 217.
45. **Loewy, A. and Van der Heide, R.,** Uber die Aufrahme des Arhylalkohols durch die Atmung, *Biochem. Z.,* 86, 125, 1918.
46. **Haggard, H. W.,** Studies on the absorption, distribution and elimination of alcohol. IX. The concentration of alcohol in the blood causing primary cardiac failure, *J. Pharmacol. Exp. Ther.,* 71, 358, 1941.
47. **Lochner, A., Cowley, R., and Brink, A. J.,** Effect of ethanol on metabolism and function of perfused rat heart, *Am. Heart J.,* 78, 770, 1969.
48. **Loomis, T. A.,** Effect of alcohol on myocardial and respiratory function. The influence of modified respiratory function on the cardiac toxicity of alcohol, *Q. J. Stud. Alcohol,* 13, 461, 1962.
49. **Regan, T. J., Koroxenidis, G., Moschos, C. B., Oldewurtel, H. A., Lehan, P. H., and Hellems, H. K.,** The acute metabolic and hemodynamic responses of the left ventricle to ethanol, *J. Clin. Invest.,* 45, 270, 1966.
50. **Goodman, L. S. and Gilman, A.,** *The Pharmacological Basis of Therapeutics,* 5th ed., Macmillan, New York, 1975.
51. **Aviado, D. M.,** The pharmacology of the pulmonary circulation, *Pharmacol. Rev.,* 12, 159, 1960.
52. **Nakano, J. and Prancan, A. V.,** Effects of adrenergic blockade on cardiovascular responses to ethanol and acetaldehyde, *Arch. Int. Pharmacodyn. Ther.,* 196, 259, 1972.
53. **Schneider, F. H.,** Acetaldehyde-induced catecholamine secretion from the cow adrenal medulla, *J. Pharmacol. Exp. Ther.,* 177, 109, 1971.
54. **Newman, W. H. and Valicenti, J. F., Jr.,** Ventricular function following acute alcohol administration. A strain-gauge analysis of depressed ventricular dynamics, *Am. Heart J.,* 81, 61, 1971.
55. **Ganz, V.,** The acute effect of alcohol on the circulation and on the oxygen metabolism of the heart, *Am. Heart J.,* 66, 494, 1963.
56. **Gould, L.,** Cardiac effects of alcohol, *Am. Heart J.,* 79, 422, 1970.
57. **Horwitz, L. O. and Atkins, J. M.,** Acute effects of ethanol on left ventricular performance, *Circulation,* 49, 124, 1974.
58. **Mendoza, L. C., Hellberg, K., Rickart, A., Tillich, G., and Bing, R. J.,** The effect of intravenous ethyl alcohol on the coronary circulation and myocardial contractility of the human and canine heart, *J. Clin. Pharmacol.,* 11, 165, 1971.
59. **Pitt, B., Sugishita, Y., Green, H. L., and Friesinger, G. C.,** Coronary hemodynamic effects of ethyl alcohol in the conscious dog, *Am. J. Physiol.,* 219, 175, 1970.
60. **Symbas, P. N., Tyras, D. H., and Balowin, B. J.,** Left ventricular function during acute ethanol intoxication and hemodialysis, *J. Surg. Res.,* 15, 207, 1973.
61. **Webb, W. R. and Degerli, I. U.,** Ethyl alcohol and the cardiovascular system: Effects on coronary blood flow, *JAMA,* 191, 1055, 1965.
62. **Wong, M.,** Depression of cardiac performance by ethanol unmasked during autonomic blockade, *Am. Heart J.,* 86, 508, 1973.

63. **Nakano, J. and Moore, S. E.,** Effect of different alcohols on the contractile force of the isolated guinea pig myocardium, *Eur. J. Pharmacol.,* 20, 266, 1972.

64. **Spann, J. F., Jr., Mason, D. T., Beisen, G. D., and Gold, H. K.,** Actions of ethanol on the contractile state of the normal and failing cat papillary muscle (abstr.), *Clin. Res.,* 16, 249, 1968.

65. **Loeb, O.,** Die Wirkung des Alkohols auf das Warmbluterherz, *Arch. Exp. Pathol. Pharmakol.,* 52, 459, 1905.

66. **Bachem, C.,** Ueber die Blutdruckwirkung kleiner Alkoholgaben bei intravenoser Injektion, *Arch. Int. Pharmacodyn. Ther.,* 14, 437, 1905.

67. **Dixon, W. E.,** Action of alcohol on circulation, *J. Physiol.* (London), 35, 346, 1907.

68. **Mason, D. T.,** Usefulness and limitations of the rate of rise of intraventricular pressure (dp/dt) in the evaluation of myocardial contractility in man, *Am. J. Cardiol.,* 23, 516, 1969.

69. **Kostis, J. B., Horstmann, E., Mavrogeorgis, E., Radzius, A., Jr., and Goodkind, M. J.,** Effects of alcohol on the ventricular fibrillation threshold in dogs, *Q. J. Stud. Alcohol,* 34, 1315, 1973.

70. **Madan, B. R. and Gupta, R. S.,** Effect of ethanol in experimental auricular and ventricular arrhythmias, *Jpn. J. Pharmacol.,* 17, 683, 1967.

71. **Paradis, R. R. and Stoelting, V.,** Conversion of acetyl-strophanthidin-induced ventricular tachycardia to sinus rhythm by ethyl alcohol, *Arch. Int. Pharmacodyn. Ther.,* 157, 312, 1965.

72. **Nakano, J. and Kessinger, J. M.,** Cardiovascular effects of ethanol, its congeners and synthetic bourbon in dogs, *Eur. J. Pharmacol.,* 17, 195, 1972.

73. **Simaan, J. A. and Aviado, D. M.,** Hemodynamic effects of aerosol propellants, *Toxicology,* 5, 127, 1975.

Part II
Hydrocarbon Propellants

Chapter 4
PROPANE

Propane is one of three hydrocarbon propellants used in aerosol products. In a classification proposed in 1974, propane was grouped together with vinyl chloride and dichlorodifluoromethane (FC 12), which are characterized as high-pressure propellants of intermediate toxicity.[1] This chapter reviews the literature and describes some experiments that will permit a comparison of the inhalational toxicity of propane with that of other hydrocarbons (Chapters 5 and 6) and with that of chlorinated solvents and alcohols (Chapters 1 to 3) used in aerosol products.

A. Review of the Literature

Propane is a paraffinic hydrocarbon that is present in natural gas to the extent of 12% by volume. Its structural formula is $CH_3CH_2CH_3$; the molecular weight is 44.09, the freezing point is -187°C, and the boiling point is -42.1°C. At a temperature of about 650°C, propane breaks down to ethylene and methane. Partial dehydrogenation of propane gives rise to propylene. Its vapor pressure is 130.3 psig at 70°C. Various chemical and physical properties are mentioned elsewhere.[2-5]

The occurrence of propane in the atmosphere is due to natural gas emissions, auto exhaust, and furnaces. The propane concentration in auto exhaust equals only about 1% that of ethylene.[6] Bida[7] found that propane is present in the atmosphere of gas refineries to the extent of 65.4 mg/m^3 at 250 m distance and 45 mg/m^3 at 1000 to 3000 m distance. The same author also found that in the autumn and winter period the concentration of propane in air is higher than in the spring and summer. Kopczynski et al.[8] found a value of 47.6 ppb of propane in the St. Louis area in December 1972. Propane is also detected in the air exhaled by healthy people,[9] and in the condensable phase of cigarette smoke to the extent of 0.7 mol %.[10] Propane can be detected in the atmosphere and in various tissues by gas chromatography.[11] The properties and safety precautions to be taken in dealing with propane and butane were discussed by Wenzel.[12]

1. Consumer Products

Bergwein[13,14] reported on the use of propane mixed with fluorocarbons, isobutane, methylene chloride, and other propellants in the manufacture of shaving creams, perfumes, dyes, automobile wax emulsion polishes, and insecticides. Propane is also contained in a disinfectant spray to the extent of 2.5 to 4.5%.[15] The efficacy of the halogenated hydrocarbons as propellants has been increased by adding propane and butane.[16] The use of propane in the aerosol industry was discussed by Larde[17] and various propellant blends containing propane have been reported on elsewhere.[18,19] However, the amount of propane that can be added to fluorocarbons is limited in quantity to render the mixture nonflammable.[20] The properties and flammability of mixtures of propane and various fluorocarbons were also discussed.

2. Industrial and Laboratory Uses

Besides its use as a fuel, propane is also used as a solvent for preparing low-pressure ethylene copolymers[21] and as a fuel in the canning industry.[22] Alone or mixed with liquid nitrogen, propane is also used for tissue preservation. The study of Epstein and O'Connor[23] on neurons quenched in liquid gases showed no significant difference in oxygen uptake of cells prepared in either liquid propane or liquid nitrogen. They concluded that propane is a more reasonable choice as it has no significant toxic effect on cells. On the other hand, Fishbein and Stowell[24] found an 80% loss in the activity of succinate-cytochrome C reductase complex isolated from mouse liver, after quenching in liquid propane. Winckler[25] found that artifacts develop within 24 hr after quenching of tissue in liquid nitrogen, due to recrystallization of the frozen tissues.

3. Human Toxicity

Ambrosio et al.[26] reported on some laboratory investigations of workers bottling commercial liquid gases (butane and propane). Most of the workers complained of respiratory symptoms, e.g., dry cough and dry throat, and gastrointestinal symptoms. Of particular interest are the electrocardiographic findings in some workers, which indicate sinus tachycardia, extrasystole, and incomplete right bundle branch block.

An attempted murder by "calor gas," which contains 5 to 18% propane, was reported in 1970 by Baldok.[27] A case of death from incomplete combustion of fuel propane gas was reported by Watanabe et al. in 1970; mortality was attributed to the formation of carbon monoxide.[28] Lactic acid production in propane poisoning was reported as slight.[29]

4. *Toxicologic Investigation in Animals*

There is limited information on the toxicity of propane in animals. Exposure to a 50% mixture of propane-butane for about ½ hr daily for 30 days resulted in slight hypochromic anemia in guinea pigs.[30]

Cardiac and pulmonary effects – Inhalation of 10% propane did not induce arrhythmia in the mouse; however, propane sensitized the heart to epinephrine-induced arrhythmia.[31] In the monkey, inhalation of 10% propane induced an insignificant decrease in myocardial force of contraction and aortic blood pressure (1.1% and 1.5%, respectively).[32] Inhalation of 20% propane aggravated these changes and caused 13.9% and 8.9% decreases in the same parameters.[32] Krantz et al. found that propane and isobutane sensitize the myocardium of the dog to epinephrine.[33] In a 10% concentration, propane caused respiratory depression in monkeys without any effect on pulmonary compliance.[34,35]

Anesthetic activity – In 1926, Brown and Henderson[36] examined the anesthetic property of propane in cats. They concluded that propane is not a useful anesthetic because the percent concentration required for relaxation of the skeletal muscles and true anesthesia is high, e.g., 89% propane failed to induce anesthesia in the cat within 4 min of inhalation; blood pressure showed a steady fall from 125 to 30 mmHg at the end of 4 min.

Miscellaneous effects – Propane, among other aliphatic gaseous hydrocarbons, exhibits an effect on the surface tension and electrical conductivity of model membranes of lipids, proteins, and mixtures of both.[37] Furthermore, propane induces swelling in mitochondria prepared with good respiratory control.[38] Propane has no effect on the color of a 2% suspension of rabbit erythrocytes in saline; however, after contact for 18 hr in closed Turner bulbs, propane changed the pH from 6.8 to 5.5.[39] It also inhibits the lysis of whole cells and cell wall fragments of *Micrococcus lysodeikticus* by lysozyme.[40] Patty and Yant[41] found the odor and physiological response augmented with increasing C atoms from propane to butane.

B. Hemodynamic Effects in Dogs

The study of the myocardial and hemodynamic effects of propane followed practically the same procedure used with chlorinated solvents (Chapter 2). The hemodynamic effects of propane in the open-chest dog preparation are summarized in Table 4.1. A record from a typical experiment is shown in Figure 4.1. In a concentration of 1%, propane elicited no significant effect on any of the parameters studied. The minimal effective concentration of propane varied from one experimental animal to another. Thus while a 2.5% concentration of propane induced various hemodynamic changes in one preparation, a 5% concentration was needed to elicit the same changes in another preparation. The threshold effective concentration averaged 3.3 ± 0.53%. However, a progressive increase in concentration of inhaled propane brought about a decrease in the various parameters depicted in Figure 4.2. These changes are as follows.

Decrease in myocardial contractility – The first significant change induced by propane is a decrease in myocardial contractility, as reflected by the decrease in maximal rate of rise of left ventricular pressure (dp/dt). This effect is brought about by as low a concentration as 2.5%. Increase of the concentration of propane in the inhaled mixture brought about a progressive and concentration-dependent decrease in myocardial contractility; thus inhalation of 2.5%, 5.0%, 10.0%, 15.0%, and 20.0% was associated with 4.5%, 7.2%, 11.0%, 14.9%, and 29.4% decreases in inotropic activity of the myocardium, respectively. The 20% concentration of propane was balanced by the appropriate volume of oxygen to prevent any hemodynamic changes due to hypoxia.

Decrease in mean aortic pressure and stroke work – At a concentration of 5%, propane induced a significant decrease in aortic pressure averaging 3.1% of the control value. This decrease in aortic pressure is gradually intensified by increasing the concentration of propane in the inhaled mixture. Thus, 10%, 15%, and 20% propane brought about decreases of 4.3%, 6.9%, and 9.3%, respectively. Concentrations of propane, amounting to 5%, 10%, 15%, and 20% brought about a decrease in stroke volume of 7.5%, 11.8%, 12.8%, and 17.4% and in stroke work of 10.3%, 15.0%, and 18.0%, and 24.2%, respectively.

Decrease in cardiac output – Propane induced not only a decrease in myocardial contractility but also a decrease in cardiac output. This latter effect

TABLE 4.1

The Myocardial and Hemodynamic Effects of Progressively Increasing Concentrations of Propane in Dogs*

	MPAP cm H_2O	MLAP cm H_2O	EMPAP cm H_2O	LVP mmHg	LVEDP mmHg	dp/dt mmHg/sec	MAP mmHg	MPAF ml/min	HR beats/min	Vascular resistance dynes·sec/cm^5 Pulmonary	Systemic	Stroke vol ml	Stroke work g·meter
Control	29.1 ± 4.1	6.7 ± 1.5	22.6 ± 3.2	150 ± 9.4	3.8 ± 1.3	3417 ± 341	126 ± 4.4	1610 ± 81	180 ± 10	818 ± 95	6012 ± 340	8.8 ± 0.7	15.6 ± 1.2
Propane 2.5%	29.4 ± 3.7 +0.3 ± 0.7 NS	6.6 ± 1.5 −0.1 ± 0.3 NS	22.8 ± 3.1 +0.3 ± 0.7 NS	149 ± 9.8 −1 ± 1.4 NS	3.8 ± 1.3 0 ± 0 NS	3333 ± 348 −84 ± 26 0.02	126 ± 4.3 0 ± 0.8 NS	1590 ± 89 −20 ± 26 NS	182 ± 9 +2 ± 2.6 NS	835 ± 89 +17 ± 18 NS	6100 ± 332 +88 ± 98 NS	8.5 ± 0.5 −0.3 ± 0.3 NS	15.1 ± 1.2 −0.5 ± 0.6 NS
Propane 5.0%	29.3 ± 3.5 +0.2 ± 0.5 NS	6.3 ± 1.4 −0.4 ± 0.3 NS	22.9 ± 3.1 +0.3 ± 0.6 NS	145 ± 9.6 −4 ± 2.0 NS	4.3 ± 1.5 +0.5 ± 0.5 NS	3188 ± 359 −229 ± 50 0.01	122 ± 4.1 −4 ± 1.0 0.02	1540 ± 107 −70 ± 30 NS	184 ± 9 +4 ± 2.8 NS	867 ± 78 +49 ± 19 0.05	6122 ± 367 +110 ± 113 NS	8.1 ± 0.6 −0.7 ± 0.3 NS	14.0 ± 1.3 −1.6 ± 0.4 0.02
Propane 10.0%	28.9 ± 3.5 −0.2 ± 0.8 NS	6.1 ± 1.2 −0.6 ± 0.3 NS	22.8 ± 3.2 +0.2 ± 0.3 NS	144 ± 9.7 −6 ± 2.6 0.05	5.0 ± 1.8 +1.2 ± 0.8 NS	3063 ± 356 −354 ± 81 0.01	121 ± 4.5 −5 ± 1.3 0.01	1480 ± 107 −130 ± 37 0.05	186 ± 9 +6 ± 3.0 NS	899 ± 97 +81 ± 12 0.01	6318 ± 420 +306 ± 157 NS	7.7 ± 0.5 −1.1 ± 0.4 0.05	13.2 ± 1.2 −2.4 ± 0.6 0.02
Propane 15.0%	27.9 ± 3.3 −1.2 ± 0.7 NS	5.5 ± 1.1 −1.2 ± 0.6 NS	22.4 ± 3.4 −0.2 ± 0.6 NS	138 ± 10.1 −12 ± 3.2 0.02	6.3 ± 2.0 +2.5 ± 0.9 0.05	2938 ± 366 −479 ± 75 0.001	117 ± 5.5 −9 ± 2.2 0.01	1440 ± 112 −170 ± 66 0.05	184 ± 9 +4 ± 1.4 0.05	913 ± 117 +95 ± 32 0.02	6213 ± 404 +201 ± 137 NS	7.6 ± 0.5 −1.2 ± 0.5 0.05	12.7 ± 1.3 −2.9 ± 1.0 0.05
Propane 20.0%	27.3 ± 3.7 −1.3 ± 1.1 NS	5.0 ± 1.1 −1.7 ± 0.9 NS	22.7 ± 3.9 +0.1 ± 1.2 NS	130 ± 10.8 −20 ± 3.4 0.001	8.8 ± 3.0 +5.0 ± 1.9 0.05	2458 ± 384 −959 ± 160 0.001	115 ± 6.9 −11 ± 3.2 0.01	1380 ± 125 −230 ± 78 0.05	185 ± 10 +5 ± 2.0 0.05	963 ± 123 +145 ± 33 0.01	6192 ± 565 +180 ± 238 NS	7.2 ± 0.5 −1.6 ± 0.6 0.05	11.8 ± 1.4 −3.8 ± 1.2 0.02

*Each group of numbers represents the mean value of seven experiments, and consists of mean response ± SEM, mean difference ± SE of difference, and the significance level.

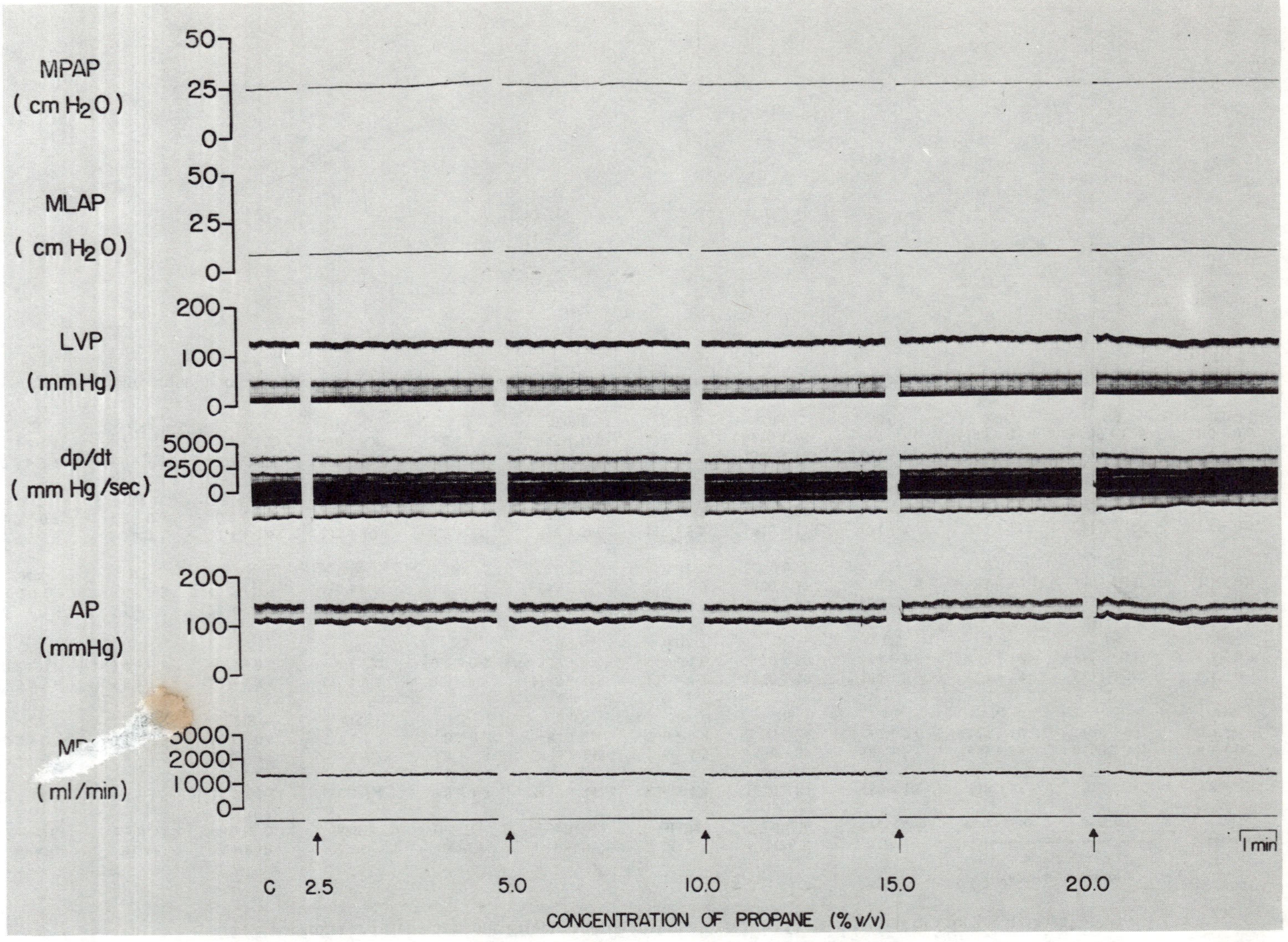

FIGURE 4.1. The effect of inhalation of progressively increasing concentrations of propane in the open-chest dog. Abbreviations are explained in the text.

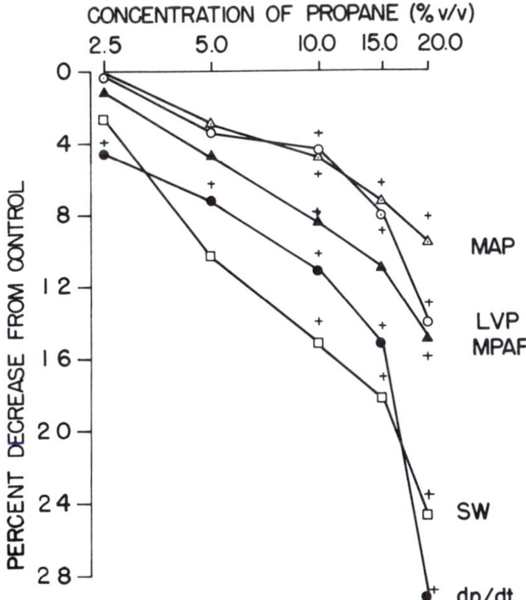

FIGURE 4.2. Mean percent decrease from control average, in response to various concentrations of propane in the open-chest dog preparation. Bars representing standard error of the mean are omitted for simplicity. + denotes significant change from control.

is due most probably to the former. A decrease in cardiac output of 4.7%, 8.4%, 10.8%, and 14.7% was induced by 5%, 10%, 15%, and 20% concentrations of propane, respectively. A decrease in left ventricular pressure of 3.4%, 4.2%, 7.9%, and 13.8% was induced by the same concentrations of propane, respectively.

Increase in pulmonary vascular resistance – A gradual increase in pulmonary vascular resistance, averaging 2.6%, 7.2%, 10.3%, 11.1%, and 17.2%, was elicited under the influence of 2.5%, 5.0%, 10%, 15%, and 20% of propane, respectively.

C. Discussion of Hemodynamic Effects

The major effects of propane are elicited on the myocardium. Thus, the first significant change in any of the parameters studied, in fact an effect that appeared at as low a concentration as 2.5%, was a decrease in myocardial contractility of 4.5%. This negative inotropic effect is concentration-dependent and is not accompanied by a negative chronotropic effect. An increase in heart rate was observed in this study. This effect is due to the stimulation of baroreceptors as a consequence of hypotension. The decrease in cardiac output, in the absence of any significant decrease in heart rate, is mainly secondary to the decrease in myocardial contractility.

The gradual and significant increase in pulmonary vascular resistance deserves a comment. A nonsignificant and slight increase in mean pulmonary arterial pressure was observed with the lower concentrations of propane. Meanwhile, a decrease in mean pulmonary arterial flow was elicited by the same concentrations of propane. Since pulmonary vascular resistance is the quotient of the effective mean pulmonary arterial pressure and pulmonary arterial flow, it appears that the increase in pulmonary vascular resistance does not necessarily reflect the actual state of the pulmonary blood vessels. The increase in pulmonary vascular resistance is a consequence of the relative changes in pulmonary blood pressure and pulmonary blood flow. Other factors affecting pulmonary circulation either directly or indirectly cannot be excluded.

D. Summary of Hemodynamic Effects

Propane, administered by inhalation to the open-chest dog preparation, induced the following myocardial and hemocynamic changes: (a) decrease in inotropism of the heart, (b) decrease in mean aortic pressure and stroke work, (c) decrease in cardiac output, and (d) increase in pulmonary vascular resistance. The minimal effective concentrations eliciting significant hemodynamic changes in a group of seven dogs averaged 3.3% v/v.

Chapter 5

BUTANE

Butane (*n*-butane) and propane are used as aerosol propellants both in the United States and Europe. In 1968, Kaempfer[42] stated that "... propane and butane are excellent solvents. They may be mixed to a very large extent with alcohol, with mineral and vegetable oils and also with chlorinated and fluorinated hydrocarbons. They are completely inert ... and ... their natural odour is hardly perceptible." The purity of the butane-propane mixture used in the aerosol industry is different from that used as fuel gas, and the butane content of the former is usually a mixture of butane and isobutane in an approximate ratio of 3:1. Butane is also used as a fuel and as a starting material for polymerization plants to convert it to iso-octane which in turn is used to increase the antiknock rating of aviation and motor fuel.[43]

A. Review of the Literature

Butane is a flammable gas that boils at -0.5°C. It is slightly soluble in water (0.15 v/v) but highly soluble in alcohol (18 v/v), ether (25 v/v) and chloroform (30 v/v). The molecular weight is 58.12.

Butane (C_4H_{10}) is a saturated aliphatic hydrocarbon occurring in natural gas. Its presence in the atmosphere is attributed to natural gas leakage, petroleum gas leakage, or diffusion through soil from petroleum deposits. Butane, resulting from radical polyethylene decomposition, has been detected in milk stored in polyethylene containers.[44]

Gordon et al.[45] have determined the average concentration of butane and isobutane in the Los Angeles atmosphere in the fall of 1967 and found values for butane ranging from 132 to 304 ppb (parts per billion parts air) and for isobutane, 44 to 74 ppb. These values varied according to the hour of sampling. Altshuller et al.[46] conducted a similar experiment and concluded that butane appears to arise from sources other than the pentanes or fuel hydrocarbons.

Inhalation toxicity in animals – In 1969, Shugaev[47] determined the concentration that kills 50% (LC_{50}) of mice and rats. Furthermore, he also determined the hydrocarbon content in various tissues by gas-liquid chromatography. The LC_{50} is 680 mg/l for mice (exposed for 2 hr) and 658 mg/l for rats (exposed for 4 hr). The effective concentration of butane, at LC_{50} level, in the mouse brain is 77.9 mg/100 g. The following concentration of butane was also found in various tissues of rat: 75.1, 49.2, 44.1, 52.2, and 208.6 mg/100 g in brain, liver, kidneys, spleen, and perinephric fat, respectively. The author also found parallelism between toxicity and effective butane concentration in the brain. This finding seems to be in contradiction to what was reported by the same author a year earlier.[48] Potentiation of the narcotic effect of butane and isobutylene was observed in 83% and an additive effect in the other 17% of experiments conducted on rats and mice.[47,49]

Neurotoxicity – The effect of butanes and pentanes on the central nervous system was studied by Stoughton and Lamson.[50] They found that butane in a concentration of 13% v/v produced light anesthesia within 25 min in mice. Only 1 min was required to induce the same effect with 22%; this concentration produced loss of posture within 15 min. Isobutane, on the other hand, in concentrations of 15%, 20%, and 23% produced light anesthesia in mice within 60, 17, and 26 min, respectively. In a concentration of 35%, however, isobutane induced loss of posture in mice in 25 min. They concluded that branched isomers of alkane series are less active and less toxic than the straight chain compounds.

Cardiotoxicity – In the course of investigating the ability of butane to sensitize the heart to the action of epinephrine in dogs, Chenoweth[51] found that only 3 out of 15 trials terminated in ventricular fibrillation.

Other biologic effects – Irradiated butane has been reported to cause eye irritation. The average threshold time for eye irritation of individuals exposed to 2 ppm of irradiated butane was 240 sec.[52] However, 3 ppm irradiated with nitrogen oxide failed to cause any sign of eye irritation; 6 ppm caused low but significant eye irritation.[53] Frommer and associates[54,55] found that butane is hydroxylated by rat liver microsomes to the isomeric alcohol.

The negative logarithm of the odor threshold concentration is 3.31 for propane and 4.28 for butane.[56] The concentration of butane that would have a noticeable odor is 5.24×10^{-5} mol/l or

3.05 mg/l. Chambers,[43] however, reported that butane itself is odorless and a small amount of stench is added to it to detect any leakage.

Butane, amongst other *n*-alkanes, has a pronounced inhibitory effect on the growth of bacteria, actinomycetes, fungi, algae, plant seeds, chick embryos, and the various states of metamorphosis of the fruit fly.[57] This effect is increased with elongation of the chain length in a series of hydrocarbons.

Butane inhibits the cell lysis of *Micrococcus lysodeikticus* by egg-white lysozyme.[58] Furthermore the release of mucopeptide from the spore coats of *Bacillus megaterun* by egg-white lysozyme was also inhibited by bubbling butane into the enzyme solution.[40] It was concluded that butane inhibits the binding of the enzyme to the bacterial cell or spore.

Ethylene, which is supposed to serve as a natural plant hormone, brings about swelling of the mitochondria prepared from heads of cauliflower (*Brassica oleracea* var. *botrytis L.*) and seedlings of Alaskan pea.

B. Hemodynamic Effects in Dogs

The hemodynamic effects of butane were examined in the anesthetized open-chest dog preparation, similar to that used for the study of methylene chloride (Chapter 2). The preparation was allowed to stabilize for a period of up to 30 min after which 0.5%, 1.0%, 2.5%, 5%, and 10% concentrations of butane in air were administered via the inlet of the respirator, for 5 min. Various concentrations were administered in an ascending order and successively. All data were analyzed by paired t-test, the criterion for significance being P less than 0.05. The results are summarized in Table 5.1, Figure 5.1 and Figure 5.2.

Decrease in mean pulmonary arterial flow and stroke volume – The inhalation of 0.5% butane brought about a significant decrease (4.1%) in cardiac output, as reflected in the measurement of pulmonary arterial blood flow. Increase in butane concentration was accompanied by a further decrease in cardiac output; thus concentrations of 1, 2.5, 5, and 10% butane were associated with reductions of 4.8, 6.0, 7.4, and 10.1%, respectively. In the absence of any significant changes in the heart rate, decrease in the cardiac output was also reflected by a decrease in stroke volume. This parameter is reduced by 2.8, 4.8, 6.3, 8.0, and 10.2% after the inhalation of 0.5, 1.0, 2.5, 5.0, and 10% concentrations of butane in air, respectively.

Decrease in myocardial contractility and stroke work – Myocardial contractility, as gauged by the rate of rise in left ventricular pressure, dp/dt, was gradually decreased by inhalation of butane. Thus 0.5% butane induced a decrease in dp/dt of 2.3% of the average control value. Increase in the butane concentration to 1.0, 2.5, 5, and 10% was associated with a progressive decrease in myocardial contractility of 4.5, 7.6, 11.2, and 17.1%, respectively. The decrease in myocardial contractility was accompanied by a greater decrease in stroke work.

Decrease in left ventricular and aortic pressure – A gradual decrease in the left ventricular pressure was observed under the influence of gradually increasing concentrations of butane. Thus a decrease of 2.3, 3.7, 5.2, 7.5, and 11.7% was induced by 0.5, 1.0, 2.5, 5.0, and 10% concentrations of butane, respectively. A decrease of 2.8, 3.0, 3.8, 5,7, and 12.4% in the aortic pressure was also induced by the above-mentioned concentrations, respectively.

No effect on vascular resistance and other parameters – Various concentrations of butane did not bring about significant changes in left atrial and pulmonary arterial pressures, and in pulmonary and systemic vascular resistances. However, at high concentrations (10%) a trend towards a decrease in the systemic vascular resistance was observed.

C. Discussion of Hemodynamic Effects

The effect of butane on the central nervous system was studied by Stroughton and Lamson[50] but its effect on the cardiovascular system has not been reported hitherto. The most significant feature of the present investigation is the decrease in cardiac output (4.1%), accompanied by a decrease in myocardial contractility (2.3%), which suggests that the former effect might be due, at least partly, to the latter action. The possibility of a decrease in venous return due to pooling of blood in capacitance vessels cannot be excluded. The effect of butane on vasculature could be a direct one, or partly induced via central depression of the vasomotor center. This effect on venous return, with the consequent decrease in filling pressure, is also shown from the decrease in stroke volume induced by butane. However, it is important to point out that there is no significant effect

TABLE 5.1

Hemodynamic Effects of Inhalation of Progressively Increasing Concentrations of Butane in Dogs*

	MPAP cm H_2O	MLAP cm H_2O	EMPAP cm H_2O	LVP mmHg	LVEDP mmHg	dp/dt mmHg/sec	MAP mmHg	MPAF ml/min	HR beats/min	Vascular resistance dynes·sec/cm^5 Pulmonary	Vascular resistance dynes·sec/cm^5 Systemic	Stroke vol ml	Stroke work g·meter
Control	23.6 ± 1.97	8.8 ± 2.16	14.9 ± 2.07	147 ± 10.1	7.6 ± 1.12	3630 ± 477.9	123 ± 9.3	1710 ± 153.6	191 ± 16.7	536 ± 102.8	5472 ± 545.2	9.1 ± 0.97	15.8 ± 1.74
0.50%	24.0 ± 2.03	8.5 ± 2.15	15.5 ± 2.22	144 ± 9.9	8.2 ± 0.92	3545 ± 468.2	120 ± 10.6	1640 ± 146.1	191 ± 17.1	578 ± 109.2	5490 ± 615.9	8.8 ± 0.94	14.9 ± 1.70
	+0.4 ± 0.54	−0.3 ± 0.15	+0.6 ± 0.52	−3 ± 1.03	+0.6 ± 0.60	−85 ± 47.2	−3 ± 1.8	−70 ± 12.2	−0.6 ± 1.43	+42 ± 19.8	+18 ± 104.1	−0.3 ± 0.11	−0.9 ± 0.15
	NS	NS	NS	0.05	NS	NS	NS	0.01	NS	NS	NS	0.05	0.01
1.0%	24.3 ± 2.16	8.5 ± 2.05	15.6 ± 2.14	142 ± 10.9	8.2 ± 0.92	3460 ± 449.3	120 ± 10.4	1630 ± 151.3	192 ± 16.8	591 ±107.5	5563 ± 626.8	8.7 ± 0.97	14.6 ± 1.87
	+0.7 ± 0.70	−0.3 ± 0.32	+0.7 ± [illegible].66	−5 ± 1.58	+0.6 ± 0.60	−170 ± 85.7	−3 ± 2.1	−80 ± 12.2	+1.0 ± 0.51	+55 ± 20.5	+91 ± 118.7	−0.4 ± 0.09	−1.2 ± 0.26
	NS	NS	NS	0.05	NS	NS	NS	0.01	NS	NS	NS	0.01	0.01
2.5%	24.2 ± 2.06	8.7 ± 1.97	15.4 ± [illegible].35	140 ± 10.4	9.2 ± 0.49	3350 ± 451.4	119 ± 10.1	1610 ± 153.6	193 ± 18.4	595 ± 116.0	5542 ± 610.9	8.5 ± 1.00	14.2 ± 1.72
	+0.6 ± 0.98	−0.1 ± 0.35	+0.5 ± [illegible].13	−7 ± 2.29	+1.6 ± 1.02	−280 ± 119.7	−4 ± 1.6	−100 ± 0.2	+2.0 ± 2.11	+59 ± 42.5	+70 ± 111.7	−0.6 ± 0.09	−1.6 ± 0.10
	NS	NS	NS	0.05	NS	0.05	0.05	0.001	NS	NS	NS	0.01	0.001
5.0%	24.2 ± 2.16	8.7 ± 2.01	15.4 ± [illegible].45	136 ± 10.4	9.8 ± 0.92	3200 ± 413.8	116 ± 9.9	1590 ± 169.1	194 ± 18.9	601 ± 116.4	5521 ± 708.7	8.5 ± 1.14	13.7 ± 1.80
	+0.6 ± 1.14	−0.1 ± 0.35	+0.5 ± [illegible].43	−11 ± 3.12	+2.2 ± 0.97	−430 ± 168.3	−7 ± 1.0	−120 ± 25.5	+3.0 ± 2.67	+65 ± 56.9	+49 ± 166.2	−0.6 ± 0.17	−2.1 ± 0.15
	NS	NS	NS	0.05	NS	0.05	0.01	0.01	NS	NS	NS	0.02	0.001
10.0%	23.4 ± 1.84	8.9 ± 1.95	14.5 ± [illegible].98	130 ± 9.9	9.8 ± 1.53	2995 ± 393.7	108 ± 10.0	1550 ± 189.7	194 ± 20.0	586 ± 104.8	5256 ± 752.8	8.3 ± 1.24	12.6 ± 2.02
	−0.2 ± 1.23	+0.1 ± 0.45	−0.4 ± [illegible].33	−17 ± 2.54	+2.2 ± 0.58	−635 ± 167.1	−15 ± 1.6	−160 ± 50.9	+3.0 ± 3.77	+50 ± 58.3	−216 ± 230.6	−0.8 ± 0.30	−3.2 ± 0.42
	NS	NS	NS	0.01	0.02	0.02	0.001	0.02	NS	NS	NS	0.05	0.001

*Each set of numbers is the mean value of 6 experiments and consists of mean response ± SEM, mean difference ± SE of difference, and the significance level.

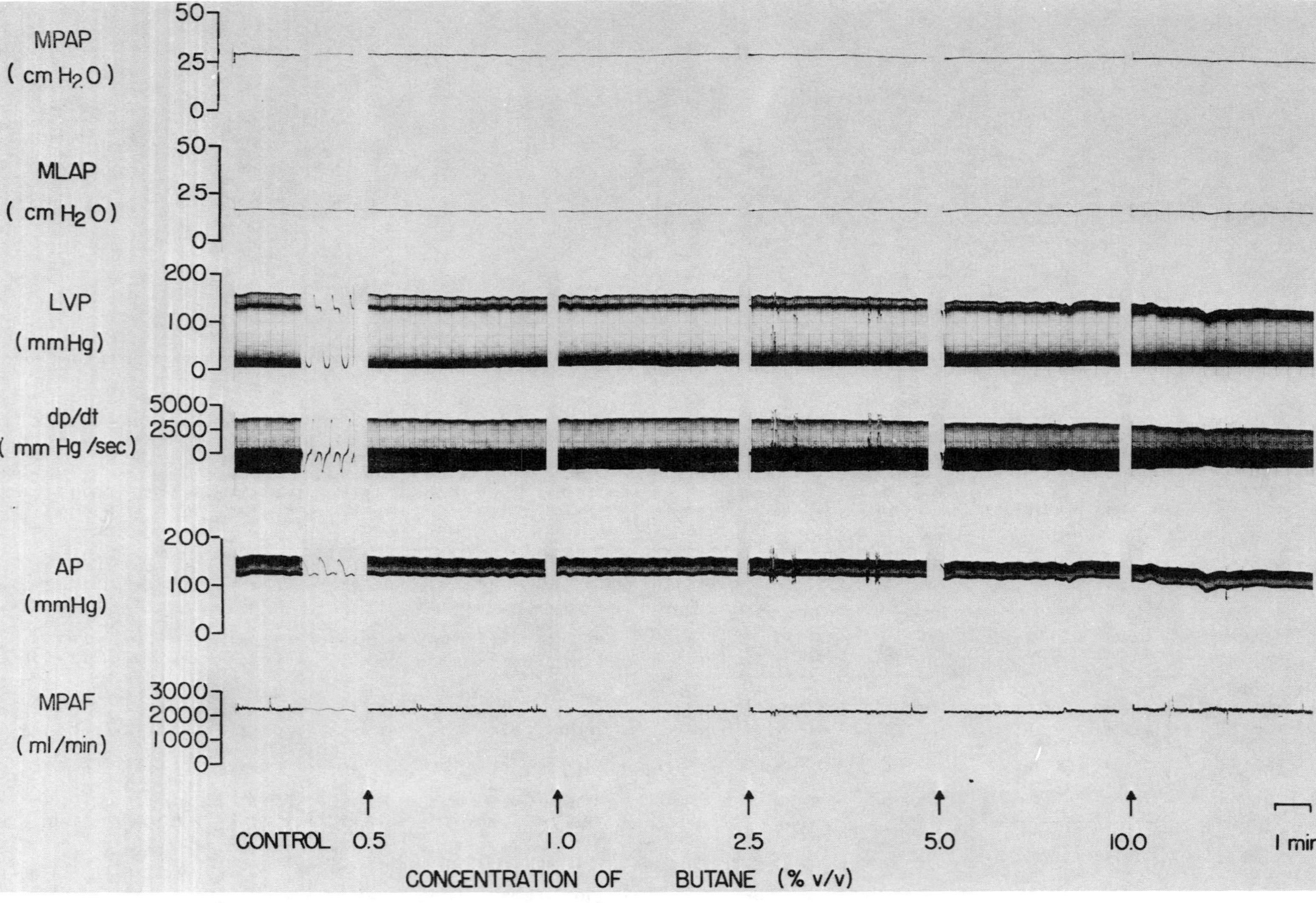

FIGURE 5.1. The effects of progressively increasing concentrations of butane on the hemodynamics of the open-chest dog preparation. MPAP = mean pulmonary arterial pressure, MLAP = mean left atrial pressure, LVP = left ventricular pressure, dp/dt = maximal rate of rise of left ventricular pressure, AP = aortic pressure, MPAF = mean pulmonary arterial flow.

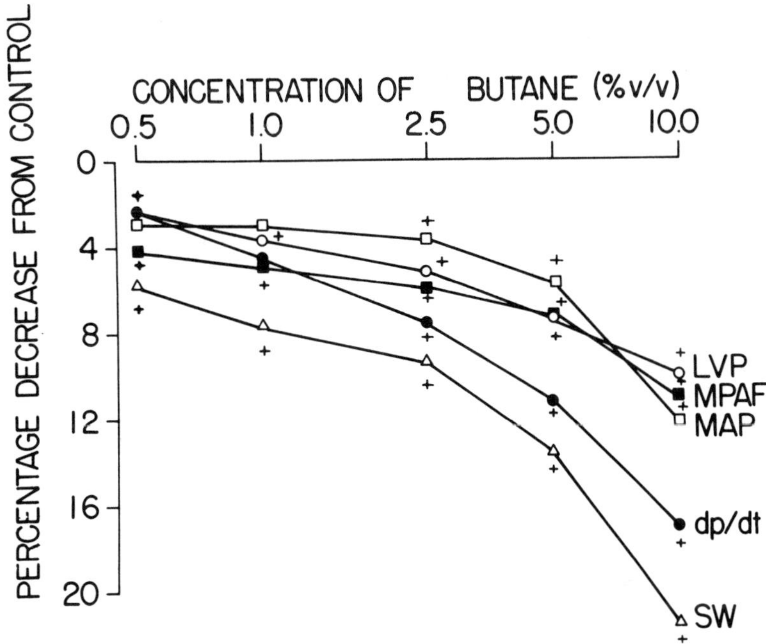

FIGURE 5.2. Concentration-response curves of butane administered by inhalational route in anesthetized open-chest dogs. LVP = left ventricular pressure, MPAF = mean pulmonary arterial flow, MAP = mean aortic pressure, dp/dt = maximal rate of rise of left ventricular pressure, SW = stroke work. + denotes significant change ($P < 0.05$).

of butane on the systemic and pulmonary blood vessels so that it appears that butane is a selective depressant of the heart.

D. Summary

The inhalation of butane in the anesthetized open-chest dog preparation brought about the following hemodynamic effects: a) decrease in cardiac output, left ventricular pressure, stroke volume, and stroke work in as low a concentration as 0.5%. b) decrease in myocardial contractility, and c) decrease in mean aortic pressure. No significant changes were observed in pulmonary arterial pressure or in vascular resistance.

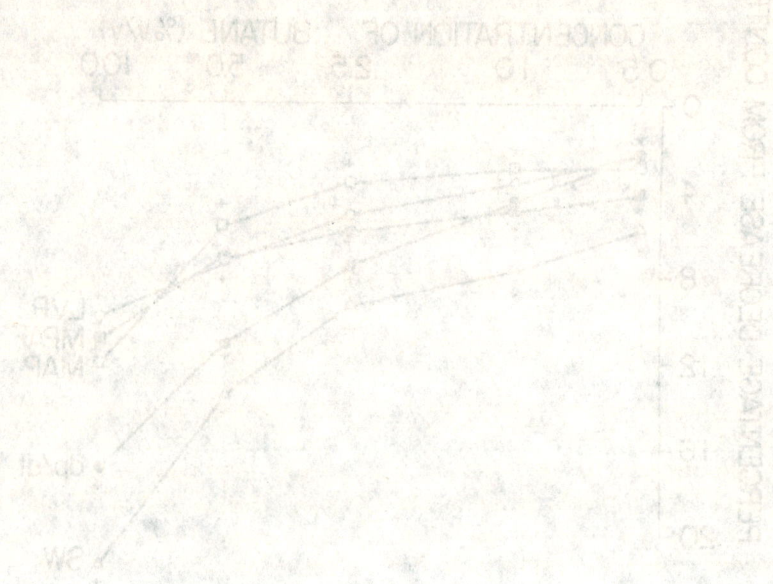

Chapter 6

ISOBUTANE

Isobutane (2-methyl propane) is the third hydrocarbon used as a propellant for aerosol products. In the classification proposed in 1974, isobutane, together with dichlorotetrafluoroethane (FC 114), belonged to the group of low-pressure propellants of intermediate toxicity.[1]

A. Review of the Literature

Isobutane is a flammable gas that melts at -138.3°C and boils at -0.5°C. Its molecular weight is 58.12. It is soluble in alcohol, ether, or chloroform.

Isobutane occurs usually mixed with propane and *n*-butane in natural gas. Baldok,[27] in 1970, reported on an attempted murder by "calor gas," which contains 5 to 18% propane, and not less than 82% of a mixture of *n*-butane and isobutane in varying amounts. The case history, as he reported, was as follows:

> A woman hid a cylinder of calor gas in her husband's bedroom. Following an argument, one morning, whilst her husband was still in bed, she left the room and then returned after allowing sufficient time for her husband to fall asleep. She closed the window and turned on the cylinder, closing the door behind her as she left. After about three hours, she found her husband still apparently asleep, and felt his pulse, waking him.

Isobutane can be detected with gas chromatographs equipped with a flame ionization detector.[6,27] The presence of isobutane in the atmosphere is attributed to natural gas leakage, petroleum gas leakage, or diesel exhaust. Battigelli[60] has determined the concentration of various gases in air samples obtained in a railroad roundhouse, 6 to 8 ft away from the exhaust of diesel engines. He found 1.4 ppm of isobutane, which does not exceed the maximal allowable concentration in air. A value of 11 ppm was found in samples taken directly over the exhausts of running engines. The average concentration of isobutane in the Los Angeles atmosphere was determined in 1967 by Altshuller et al.[46] who found that the concentration of isobutane in air is almost 25% of that of *n*-butane. Gordon et al.[45] found that isobutane occurs in from 44 to 74 ppb in air in Los Angeles. Isobutane and *n*-butane are also detected in cigarette smoke in a concentration of 0.001 and 0.006% v/v, respectively.[61]

Isobutane is used not only as a lighter fuel, but also as a propellant in several household products, such as hair spray (five products), arts and paint spray (four products), deodorant spray, color spray decorator, antirust spray, and insecticide aerosol (two products of each), room deodorant, coating spray, foam bathroom cleaner, shoe spray, shaving foam, rug shampoo and spot remover (one product of each).[62] Bergwein[14] reported on a water-containing spray product containing 76% isobutane and 24% propane.

Narcotic activity – In concentrations of 15%, 20%, and 23% in air, isobutane induced light anesthesia in mice within 60, 17, and 26 min, respectively. However, it induced loss of posture in mice inhaling 35% in 25 min.[50] Using houseflies, Bradbury and O'Carroll[63] found that the partial saturation of *n*-butane and isobutane that produces the same average knock-down time as chloroform at a partial saturation of 0.005, are 0.088 and 0.052, respectively.

Cardio- and pulmonary toxicity – Isobutane sensitizes the heart to epinephrine-induced arrhythmia in the mouse.[31] Twenty-seven percent isobutane produced apnea after 8.7 min and respiratory arrest after 15 min in anesthetized rats.[64]

Other biological effects – Isobutane inhibits the cell lysis of *Micrococcus lysodeikticus* by egg-white lysozyme.[65] *Mycobacterium phlei* is capable of growth on isobutane as its sole carbon and energy source.[66] Isobutane is hydroxylated by rat liver microsomes to the isomeric alcohol.[55]

B. Inhalational Toxicity in Mice

To determine the inhalational LC_{50}, male mice (CF-1 strain, Charles River Laboratories, Wilmington, Massachusetts) weighing 20 to 25 g were divided into six groups, each group with ten mice. The mice were then introduced into a 10-liter glass chamber and were allowed to breathe isobutane in air for 120 min; the gaseous mixture was flushed into the chamber at a rate of 3 l/min. Since high concentrations of isobutane in air were used, oxygen (25% of the volume of isobutane) was added to the gaseous mixture to prevent any death due to hypoxia. Results were then calculated according to the probit method.[67]

Table 6.1 summarizes the results of 120-min

TABLE 6.1

Acute Inhalational Toxicity of Isobutane in Mice after 120 min of Exposure*

Group	Concentration % v/v	% Mortality	LC_{50} 120 min	Regression coefficient
1	36	0		
2	40	10		
3	50	30	52.04	
4	55	50	±	0.9662
5	60	90	3.26%	
6	65	100		

*Each group consisted of ten mice.

exposures of mice to various concentrations of isobutane in air. The highest concentration that elicited no mortality was 36% v/v, while the lowest concentration that was associated with 100% mortality was 65%. The LC_{50} (95% fiducial limits) was 52.04 ± 3.26% v/v (i.e., 520,400 ppm). Mice showed signs of central nervous system depression, rapid and shallow respiration, loss of posture, and apnea.

C. Hemodynamic Effects in Dogs

The experiments reported in this section were designed to identify the cardiovascular effects of isobutane in the anesthetized dogs, as well as to find any evidence for interaction with any of the previously investigated solvents. The methods are described in Chapter 2.

Each of seven preparations was allowed to stabilize for a period of up to 30 min, after which 0.5%, 1.0%, 2.5%, 5%, and 10% concentrations of isobutane in air were administered via the inlet of the respirator, for 5 min. A period of approximately 10 min elapsed before the administration of the subsequent concentration, to allow for recovery of the preparation.

In each preparation, the concentration of isobutane that produced minimal cardiovascular effects was determined. The minimal effective concentrations of each of methyl chloroform,[68] trichloroethylene,[68] methyl ethyl ketone,[69] and methyl isobutyl ketone[69] previously determined in a similar preparation, viz., 0.1%, 0.05%, 0.09%, and 0.08%, respectively, were administered to the animals for 5 min, either separately or mixed with the threshold concentration of isobutane. The sequence of administration of the threshold concentration of isobutane, methyl chloroform, trichloroethylene, methyl ethyl ketone, and methyl isobutyl ketone, and a mixture of the threshold concentration of isobutane and various compounds, was alternated from one preparation to the other. All data were analyzed by paired comparison and Student t-test, the criterion for significance being P less than 0.05.

Effects of inhalation of various concentrations of isobutane – The effect of various concentrations of isobutane on the cardiovascular system is summarized in Table 6.2. A record from a typical experiment is shown in Figure 6.1. In concentrations of 0.5% and 1.0%, isobutane produced no significant change in any of the studied parameters. However, in a concentration of 2.5% isobutane brought about a decrease in the myocardial contractility, as shown by the decrease in maximal rate of rise of the left ventricular pressure (dp/dt), a decrease in cardiac output, stroke volume, and stroke work, averaging to 13.1%, 5.3%, 6.1%, and 9.8%, as compared with the average control, respectively. Increasing the concentration of the inhaled mixture of isobutane in air to 5% not only aggravates the above-mentioned decreases in various parameters but also decreases the left ventricular and the aortic pressures by 6.7% and 7.8%, respectively. In a 10% concentration, isobutane brought about a significant decrease in left ventricular pressure, left ventricular dp/dt, mean arterial pressure, mean pulmonary arterial flow, stroke volume, and stroke work averaging to 14.4%, 19.8%, 15.0%, 12.3%, 12.6%, and 25.6%, respectively. Furthermore, a significant increase in pulmonary vascular resistance of 16.6% was noticed with the highest concentration used, namely 10%. It is evident from these figures that the most pronounced effect of isobutane is on the myocardial contractility as reflected by the decrease both in the

TABLE 6.2

The Hemodynamic Effects of Progressively Increasing Concentrations of Isobutane in Dogs*

	MPAP cm H_2O		MLAP cm H_2O		EMPAP cm H_2O		LVP mmHg		LVEDP mmHg		dp/dt mmHg/sec		MAP mmHg		MPAF ml/min		HR beats/min		Vascular resistance dynes·sec/cm^5 Pulmonary		Vascular resistance dynes·sec/cm^5 Systemic		Stroke vol ml		Stroke work g·meter	
	C	E	C	E	C	E	C	E	C	E	C	E	C	E	C	E	C	E	C	E	C	E	C	E	C	E
Isobutane 0.5%	50.3	49.5	6.8	7.1	43.3	42.8	140	139	3.0	3.4	4521	4500	126	123	2358	2325	168	166	1259	1284	4387	4341	14.6	14.5	25.7	24.7
	−0.8 ± 0.5		−0.3 ± 1		−0.5 ± 0.3		−1 ± 2.9		+0.4 ± 0.4		−21 ± 93		−3 ± 2.3		−33 ± 40		−2 ± 1.8		+25 ± 36		−46 ± 61		−0.1 ± 0.3		−1 ± 0.9	
	NS		NS		NS		NS		NS		NS		NS		NS		NS		NS		NS		NS		NS	
Isobutane 1.0%	48.6	48.5	6.5	6.0	41.7	41.9	141	139	1.3	3.3	4313	4294	123	122	2283	2203	165	168	1243	1240	4476	4365	14.3	13.9	24.8	24.0
	−0.1 ± 1.1		−0.5 ± 0.2		+0.2 ± 1.0		−2 ± 2.7		+2 ± 0.8		−19 ± 137		−0.7 ± 1.9		−80 ± 40		+3 ± 2.1		−3 ± 25		−111 ± 125		−0.4 ± 0.4		−0.8 ± 0.5	
	NS		NS		NS		NS		NS		NS		NS		NS		NS		NS		NS		NS		NS	
Isobutane 2.5%	49.0	47.4	6.3	6.8	42.4	40.3	146	141	2.5	4.2	4458	3875	121	116	2233	2108	166	167	1246	1245	4401	4414	13.9	13.0	24.1	21.5
	−1.6 ± 0.5		+0.5 ± 0.6		−2.1 ± 0.8		−5 ± 4		+1.7 ± 1.2		−583 ± 173		−5 ±1.5		−125 ± 31		+1 ± 0.7		−1 ± 23		+13 ± 36		−0.9 ± 0.3		−2.6 ± 0.8	
	NS		NS		NS		NS		NS		NS		NS		0.01		NS		NS		NS		0.02		0.02	
Isobutane 5.0%	47.2	46.4	7.0	7.1	39.7	38.9	146	136	1.7	1.7	4354	3771	123	114	2175	2042	165	169	1142	1185	4566	4741	13.4	12.3	23.2	19.8
	−0.8 ± 0.5		+0.1 ± 0.1		−0.8 ± 0.5		−10 ± 1.7		0 ± 0		−583 ± 53		−9 ± 2.3		−133 ± 40		+4 ± 1.5		+43 ± 46		+175 ± 218		−1.1 ± 0.3		−3.4 ± 0.6	
	NS		NS		NS		0.01		NS		0.001		0.01		0.02		NS		NS		NS		0.02		0.01	
Isobutane 10.0%	43.5	42.3	7.7	7.2	35.8	35.2	142	121	2.9	2.7	4177	3365	119	101	2138	1875	164	164	1019	1185	4418	4303	13.3	11.7	22.6	16.8
	−1.2 ± 0.5		−0.5 ± 0.4		−0.6 ± 0.9		−21 ± 3.9		−0.2 ± 0.5		−8.12 ± 95		−18 ± 2.6		−263 ± 41		0 ± 0.9		+166 ± 21		−115 ± 99		−1.6 ± 0.3		−5.8 ± 0.8	
	NS		NS		NS		0.01		NS		0.001		0.001		0.01		NS		0.01		NS		0.01		0.01	

*Each group of numbers represents the mean value of seven experiments, and consists of mean control, mean response, mean difference ±SE of difference, and the significance level.

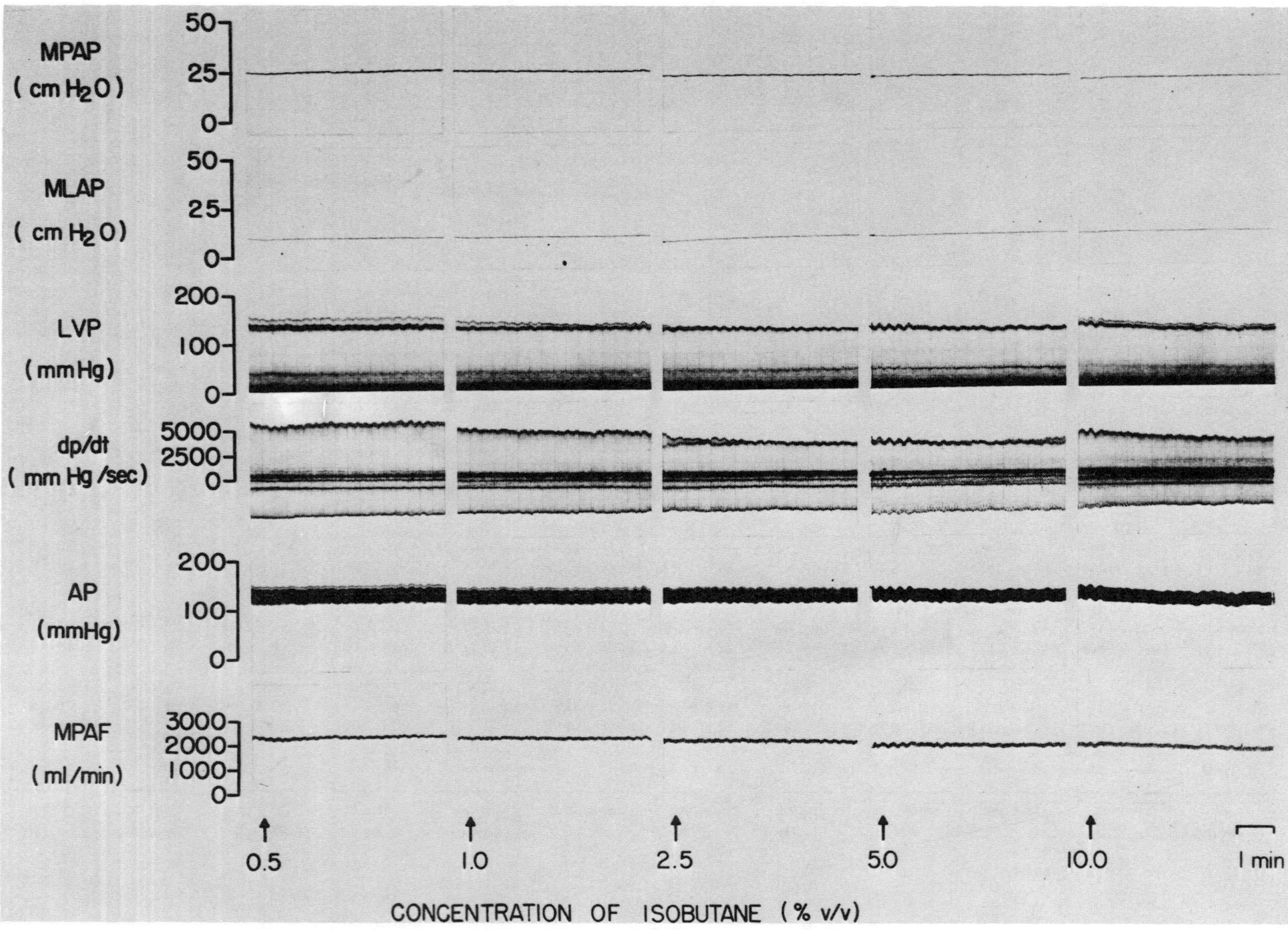

FIGURE 6.1. The effects of inhalation of progressively increasing concentrations of isobutane in the open-chest dog preparation. Abbreviations are explained in the text.

maximum rate of rise of left ventricular pressure and the stroke work. These effects are depicted in Figure 6.2.

No significant change was observed in mean pulmonary arterial pressure, mean left atrial pressure, effective mean pulmonary arterial pressure, left ventricular end-diastolic pressure, heart rate, and systemic vascular resistance.

Effects of inhalation of isobutane and halogenated hydrocarbons or ketones – The threshold effective concentrations of halogenated solvents (methyl chloroform MC, and trichloroethylene TCE),[68] and of ketones (methyl ethyl ketone MEK, and methyl isobutyl ketone MIK)[69] were previously determined in this laboratory in a similar dog preparation. In the present investigation, it was felt necessary to study the effect of these compounds alone and mixed with isobutane especially because most, if not all, of the consumer products examined contain these solvents mixed with isobutane as a propellant.

The minimal effective concentration of isobutane was determined in each of the preparations studied; its value varied between 1.0% and 2.5%, averaging 2.0% ± 0.32. In this concentration, isobutane brought about a decrease in left ventricular pressure, left ventricular dp/dt, mean pulmonary arterial flow, stroke volume, and stroke work, averaging to 5.6%, 10.4%, 5.9%, 7.0%, and 10.6%, respectively. Inhalation of 0.1% methyl chloroform brought about a decrease of 3.4%, 7.2%, 4.5%, 4.2%, 4.3%, and 8.7% in left ventricular pressure, left ventricular dp/dt, mean arterial pressure, mean pulmonary arterial flow, stroke volume, and stroke work, respectively. Trichloroethylene, on the other hand, inhaled in a concentration of 0.05%, elicited a decrease of 7.5% in left ventricular dp/dt, and of 3.8%, 4.9%, and 4.0% in mean pulmonary arterial flow, stroke volume, and stroke work, respectively.

Inhalation of a mixture of 2% isobutane and 0.1% methyl chloroform brought about a decrease of 6.8%, 6.4%, 10.5%, 6.5%, 7.9%, and 13.9% in mean arterial pressure, left ventricular pressure, left ventricular dp/dt, mean pulmonary arterial flow, stroke volume, and stroke work, respectively. A mixture of a threshold concentration of isobutane and a threshold concentration of trichloroethylene brought about a decrease in left ventricular pressure, left ventricular dp/dt, mean

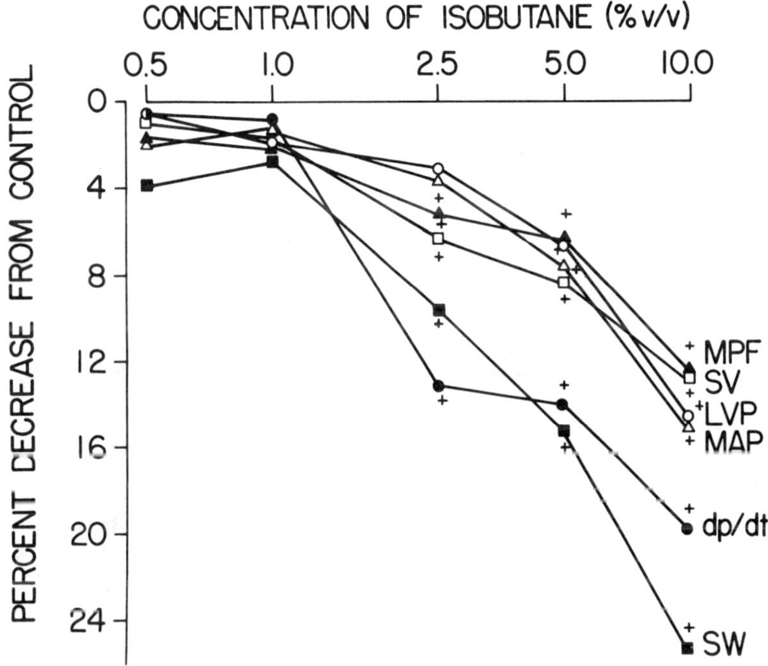

FIGURE 6.2. Mean percent decrease from control average in response to various concentrations of isobutane in open-chest dog preparation. Bars representing standard error of the mean are omitted for simplicity. + denotes significant difference from control.

pulmonary arterial flow, stroke work, and stroke volume, averaging 6.6%, 9.1%, 4.1%, 6.3%, and 8.4%, respectively. The Student t-test revealed no significant difference ($P < 0.05$) between the effects of isobutane alone or mixed with methylchloroform or trichloroethylene. These results are depicted in Figure 6.3 and summarized in Table 6.3.

In a previous study, ketones (MEK and MIK), especially in suprathreshold concentrations, caused a unique effect on the pulmonary circulation, namely, increase in pulmonary pressure and resistance. Isobutane (2%), on the other hand, decreased pulmonary arterial pressure and increased pulmonary vascular resistance mainly because of decreased pulmonary arterial flow (Figure 6.4). No significant difference could be observed between the effect of isobutane alone or mixed with the threshold concentrations of ketones (Table 6.4).

D. Discussion of Hemodynamic Effects

The present investigation was mainly concerned with the hemodynamic effects of various concentrations of isobutane in the intact, anesthetized, open-chest dog preparation. No significant changes were observed with concentrations of 0.5% and 1.0% of isobutane. However, at a concentration of 2.5%, significant decreases in myocardial contractility, mean pulmonary arterial flow, stroke volume, and stroke work were observed. These effects gradually intensified with increase in the concentration to 10%. In a concentration of 5% or 10%, furthermore, isobutane decreased the left ventricular and mean aortic pressures. A significant increase in the pulmonary vascular resistance was observed only with a 10% concentration. The results of this study demonstrate that the most predominant cardiovascular effect of isobutane is its ability to cause depression of myocardial contractility and output as shown from the decrease in the maximum rate of rise of left ventricular pressure (dp/dt) and pulmonary flow. The cardiac negative inotropic effect of isobutane is not associated with a similar negative chronotropic effect, as the heart rate is not significantly

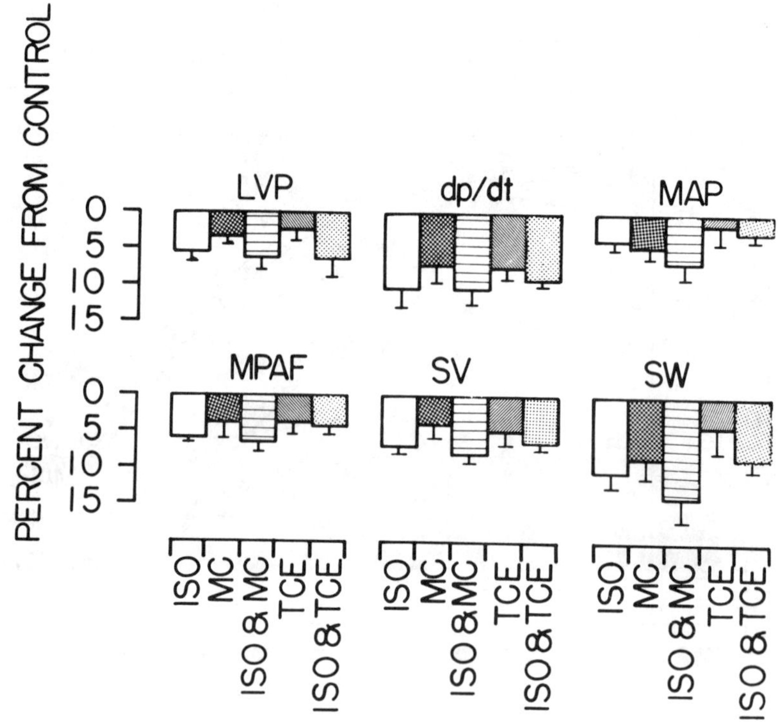

FIGURE 6.3. Effects of the threshold concentration of isobutane (ISO), methyl chloroform (MC), and trichloroethylene (TCE), and a mixture of ISO and MC or TCE, on various hemodynamic parameters. Abbreviations are explained in the text.

TABLE 6.3

The Effect of Minimal Effective Concentrations of Isobutane, Methyl Chloroform, and Trichloroethylene on the Open-chest Dog*

	MPAP cm H_2O		MLAP cm H_2O		EMPAP cm H_2O		LVP mmHg		LVEDP mmHg		dp/dt mmHg/sec		MAP mmHg		MPAF ml/min		HR beats/min		Vascular resistance dynes·sec/cm^5				Stroke vol ml		Stroke work g·meter	
																			Pulmonary		Systemic					
	C	E	C	E	C	E	C	E	C	E	C	E	C	E	C	E	C	E	C	E	C	E	C	E	C	E
Isobutane 2.0%	51.3	49.6	6.5	6.3	44.4	43.1	150	142	2.9	4.2	4646	4167	124	119	2358	2217	167	169	1225	1258	4231	4283	14.6	13.5	25.6	22.8
	−1.7 ± 0.9		−0.2 ± 0.3		−1.3 ± 0.9		−8 ± 1.8		+1.3 ± 1.3		−479 ± 113		−5 ± 1.5		−141 ± 20		+2 ± 0.9		+33 ± 15		+52 ± 46		−1.1 ± 0.2		−2.8 ± 0.7	
	NS		NS		NS		0.01		NS		0.02		0.05		0.001		NS		NS		NS		0.01		0.02	
Methyl chloroform (MC) 0.05%	38.2	37.6	7.6	6.7	30.3	30.3	139	134	3.3	3.2	4042	3782	119	114	1954	1871	159	160	960	1017	4891	4871	12.4	11.9	20.9	19.2
	−0.6 ± 1.0		−0.9 ± 0.4		0 ± 1.0		−5 ± 1.1		−0.1 ± 0.2		−260 ± 69		−5 ± 1.7		−83 ± 36		+1.0 ± 0.5		+57 ± 53		−20 ± 78		−0.5 ± 0.2		−1.7 ± 0.5	
	NS		NS		NS		0.01		NS		0.02		0.02		NS		NS		NS		NS		0.05		0.02	
Trichloro-ethylene (TCE) 0.05%	43.0	39.7	7.4	7.4	35.3	31.8	139	135	2.3	2.3	4021	3744	115	115	1863	1796	159	162	1195	1147	5048	5247	11.9	11.3	19.5	18.8
	−3.3 ± 2.3		0 ± 0.4		−3.5 ± 2.3		−4 ± 1.9		0 + 0.7		−277 ± 48		0 ± 2.5		−77 ± 25		+3 ± 1.3		−48 ± 82		+198 ± 96		−0.6 ± 0.2		−0.7 ± 0.6	
	NS		NS		NS		NS		NS		0.01		NS		0.05		NS		NS		NS		0.05		NS	
Isobutane + 0.1% MC	45.8	44.6	6.5	6.5	39.3	37.9	140	131	2.5	1.9	4125	3729	118	109	1988	1863	159	161	1219	1275	4729	4622	12.8	11.8	21.4	18.5
	−1.2 ± 0.7		0 ± 0.2		−1.4 ± 0.6		−9 ± 2.1		−0.6 ± 0.6		−396 ± 38		−9 ± 2.8		−125 ± 26		+2 ± 0.7		+56 ± 46		−107 ± 186		−1 ± 0.2		−2.9 ± 0.7	
	NS		NS		NS		0.01		NS		0.001		0.05		0.01		0.05		NS		NS		0.01		0.01	
Isobutane + 0.05% TCE	43.4	41.9	7.3	6.5	35.3	34.3	137	128	3.0	4.0	3875	3529	117	114	1829	1758	159	162	1206	1236	5168	5212	11.7	11.0	19.4	17.9
	−1.5 ± 0.6		−0.7 ± 0.4		−1 ± 0.7		−9 ± 3.6		+1 ±0.6		−346 ± 48		−3 ±1		−71 ± 12		+3 ± 0.7		+30 ± 27		+44 ± 65		−0.7 ± 0.1		−1.5 ± 0.2	
	NS		NS		NS		0.05		NS		0.001		0.05		0.01		0.01		NS		NS		0.001		0.001	

*Each set of numbers consists of mean control (C), mean experimental (E), mean difference ±SE of difference and the P value. NS = nonsignificant (i.e., P > 0.05). Abbreviations are explained in the text.

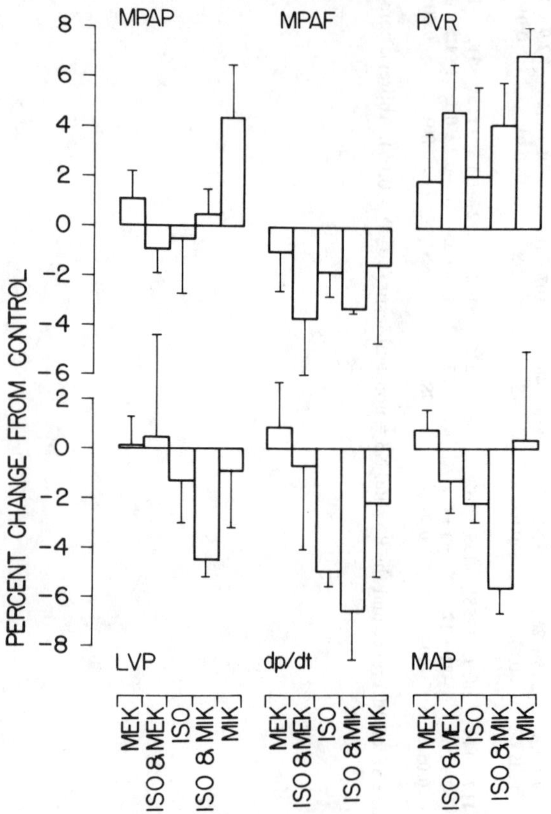

FIGURE 6.4. Effects of the threshold concentration of isobutane (ISO), methyl ethyl ketone (MEK), methyl isobutyl ketone (MIK), and a mixture of ISO and MEK or MIK. Abbreviations are explained in the text.

changed. The parameters that were significantly altered by a concentration of 2.5% of isobutane were the myocardial contractility and the pulmonary artery blood flow. At this concentration, the significant decrease in cardiac output appears to be due mainly to the attenuation of myocardial inotropic activity, since the heart rate was not significantly changed. At higher concentrations, however, pooling of blood in the capacitance vessels with the consequent decrease in venous return might participate in decreasing cardiac output. That this mechanism is involved is suggested in the light of the decrease in systemic vascular resistance elicited with the concentration of 10%.

The nonsignificant increase in pulmonary vascular resistance elicited by 5.0% concentrations of isobutane can be explained as follows: with these concentrations, no significant change in pulmonary arterial pressure, left atrial pressure, or effective mean pulmonary arterial pressure was observed. The pulmonary vascular resistance is the quotient of the effective mean pulmonary arterial pressure divided by cardiac output. Since there was no significant change in the effective mean pulmonary arterial pressure, when the cardiac output decreased, the pulmonary vascular resistance increased. Nonetheless, the significant increase in pulmonary vascular resistance observed with 10% concentration of isobutane, and averaging to 16.6% of the mean control value, seems to be due to the involvement of other components in the pulmonary circulation. The exact mechanism of action has to await further exploration.

Inhalation of threshold concentrations of isobutane, methyl chloroform, and trichloroethylene separately revealed that these three compounds share the property of depressing the myocardial contractility. Nevertheless, administration of the minimal effective concentration of isobutane mixed with the threshold concentration of trichloroethylene or methyl chloroform showed that there is no addition, potentiation, or synergism of the action of these agents. Also, no significant interaction could be detected between isobutane and ketones. This might imply a different basic mechanism of action of these substances on the cardiovascular system.

E. Summary

Isobutane exerts the following pharmacologic profile in the open-chest dog preparation: a) decrease in cardiac output, stroke volume, and stroke work; b) decrease in myocardial contractility; and c) decrease in left ventricular and aortic pressures. The threshold effective concentration of isobutane is 2.0%. No interaction could be observed between the threshold concentration of isobutane and those of methyl chloroform, trichloroethylene, methyl ethyl ketone, and methyl isobutyl ketone.

TABLE 6.4

The Effects of Minimal Effective Concentrations of Isobutane, Methyl Ethyl Ketone, and Methyl Isobutyl Ketone on the Open-chest Dog*

	MPAP cm H_2O		MLAP cm H_2O		EMPAP cm H_2O		LVP mmHg		LVEDP mmHg		dp/dt mmHg/sec		MAP mmHg		MPAF ml/min		HR beats/min		Vascular resistance dynes·sec/cm^5				Stroke vol ml		Stroke work g·meter	
																			Pulmonary		Systemic					
	C	E	C	E	C	E	C	E	C	E	C	E	C	E	C	E	C	E	C	E	C	E	C	E	C	E
Isobutane 2%	35.0	34.8	7.8	7.6	27.2	27.2	125	123	5.5	6.3	3656	3469	109	107	1438	1413	177	179	1112	1132	6076	5950	8.3	8.0	12.9	12.2
	−0.2 ± 0.8		−0.2 ± 0.2		0 ± 0.9		−2 ± 2.2		+0.8 ± 0.8		−187 ± 36		−3 ± 1.0		−25 ± 14		−2 ± 0.9		+20 ± 42		−126 ± 46		−0.3 ± 0.1		−0.7 ± 0.2	
	NS		NS		NS		NS		NS		0.02		NS		NS		NS		NS		NS		0.05		0.05	
MEK 0.09%	32.3	32.6	5.2	5.5	27.0	27.1	161	161	3.1	3.8	4531	4563	139	140	1420	1410	173	174	1193	1204	8167	8332	8.3	8.2	16.1	16.3
	+0.3 ± 0.3		±0.3 ± 0.3		+0.1 ± 0.5		0 ± 2		+0.7 ± 0.6		+32 ± 79		+1 ± 1		−10 ± 19		+1.9 ± 0.9		+11 ± 21		+165 ± 183		−0.1 ± 0.1		+0.2 ± 0.3	
	NS		NS		NS		NS		NS		NS		NS		NS		NS		NS		NS		NS		NS	
MIK 0.08%	33.5	35.0	7.2	7.2	26.3	27.8	132	131	8	8	4156	4094	114	114	1413	1388	179	181	1102	1176	6398	6463	8.0	7.8	13.1	12.7
	+1.5 ± 0.8		0 ± 0		+1.5 ± 0.8		−1 ± 2.7		0 ± 0		−62 ± 108		0 ± 3.5		−25 ± 43		+2 ± 1.9		+74 ± 9.9		+65 ± 119		−0.2 ± 0.3		−0.4 ± 0.8	
	NS		NS		NS		NS		NS		NS		NS		NS		NS		0.01		NS		NS		NS	
Isobutane + MEK	31.0	30.8	4.5	3.7	26.5	27.0	161	162	3.8	5.6	4344	4281	141	139	1450	1410	167	168	1109	1167	7819	8013	8.7	8.4	17.6	16.7
	−0.2 ± 0.3		−0.8 ± 0.5		+0.5 ± 0.7		+1 ± 6.9		+1.8 ± 1.9		−63 ± 149		−2 ± 1.8		−40 ± 25		+1 ± 0.4		+58 ± 20		+194 ± 361		−0.3 ± 0.1		−0.9 ± 0.4	
	NS		NS		NS		NS		NS		NS		NS		NS		NS		0.05		NS		NS		NS	
Isobutane + MIK	35.5	35.7	7.4	7.4	28.2	28.3	131	125	1.5	1.5	3813	3563	115	108	1513	1463	180	181	1087	1130	5941	5802	8.5	8.2	13.9	13.0
	+0.2 ± 0.3		0 ± 0		+0.1 ± 0.3		−6 ± 1.2		0 ± 0		−250 ± 88		−6 ± 0.8		−50 ± 10		+1 ± 1.6		+43 ± 12		−139 ± 55		−0.3 ± 0.1		−0.9 ± 0.3	
	NS		NS		NS		0.02		NS		0.05		0.01		0.01		NS		0.05		NS		0.05		0.05	

*Each set of numbers is calculated from four preparations and consists of mean control (C), mean experimental (E), mean difference ± SE of difference, and the P value. NS = non-significant (i.e., P > 0.05). Abbreviations are explained in the text.

REFERENCES – Hydrocarbon Propellants

1. **Aviado, D. M.,** Toxicity of propellants, in *Progress in Drug Research,* Vol. 18, Jucker, E., Ed., Birkhauser Verlag, Basel, 1974, 365.
2. **Denny, L. C. and Wickstrom, H. W., Eds.,** *Handbook of Butane-Propane Gases,* Jenkins, Los Angeles, 1947.
3. **Finley, G. H., Ed.,** *Handbook of Butane-Propane Gases,* 2nd ed., Western Gas Co., Los Angeles, 1935.
4. **Graf, L. and Hlinyansky, I.,** *Propane-Butane,* Muszaki Kiado, Budapest, 1967.
5. **Ludwig, E. H. B.,** Propane, butane and pentane, the "liquid gases," *Z. Osterr. Ver. Gas Wasserfachmaennern,* 74, 121, 1934.
6. **Altshuller, A. P. and Bellar, T. A.,** Gas chromatographic analysis of hydrocarbons in the Los Angeles atmosphere, *J. Air Pollut. Control Assoc.,* 13, 81, 1963.
7. **Bida, M. P.,** Pollution of the atmosphere with discharges from a gas refinery, *Faktory Vnesh. Sredy Ikh Znach. Zdorovya Naseleniya,* 2, 170, 1970.
8. **Kopczynski, S. L., Lonneman, W. A., Winfield, T., and Seila, R.,** Gaseous pollutants in St. Louis and other cities, *J. Air Pollut. Control Assoc.,* 25, 251, 1975.
9. **Nefedov, Y. G., Savina, V. P., Sokolov, N. L., and Ryzhkora, V. E.,** Study of contaminants in the air exhaled by man, *Chem. Abstr.,* 73, 206, 1970.
10. **Osborne, J. S., Adamek, S., and Hobbs, M. E.,** Some components of gas phase of cigarette smoke, *Anal. Chem.,* 28, 211, 1956.
11. **Nagata, T., Kojima, T., and Makisumi, S.,** Quantitative gas chromatographic determination of propane and LP (liquefied petroleum) gas in biological materials, *Jpn. J. Legal Med.,* 25, 439, 1971.
12. **Wenzel, L. R.,** Safety precautions when installing propane-air mixtures, *Gas,* 23, 44, 1947.
13. **Bergwein, K.,** Mixtures of aerosol propellants used in the U.S.A., *Seifen Oele Felte Wachse,* 93, 95, 1967.
14. **Bergwein, K.,** Pure hydrocarbons, liquefied propellants, and vinyl chloride as aerosol propellants and mixtures, *Kosmet. Parfum Drogen Rundsch.,* 17, 20, 1970, (in German).
15. **French, F. R. and Paige, J. E.,** Disinfectant spray, German patent 2,401,112, 1974.
16. **Kuz'menko, I. E., Leinasare, D., Lev, F. B., Musalyamov, F. U., and Monakhov, V. T.,** Propellant for sprays, Russian patent 321,091, 1973.
17. **Larde, R. and Couillaud, A.,** Aerosol propellant, French patent 448,203, 1966.
18. **Reed, A. B.,** Specialty propellant blends: their evaluation and application to aerosols, Chem. Spec. Manuf. Assoc., Proc. Mid-Year Meet., 48, 56, 1962.
19. **Stamm, A. and Hug, R.,** Aerosol propellants. III. Inflammable liquid propellants and methylene chloride, *Labo Pharma Probl. Tech.,* 22, 839, 1974.
20. **Scott, R. J. and Feathers, R. E.,** Aerosol propellant blends, *Soap Chem. Spec.,* 38, 154, 1962.
21. **Baxmann, F., Dietrich, J., Frese, A., and Hahman, O.,** Low-pressure ethylene copolymers, German patent 2,206, 429, 1973.
22. **Hviding, J.,** Use of propane in the canning industry, *Tidsskr. Hermetikind.,* 53, 249, 1967.
23. **Epstein, M. H. and O'Connor, J. S.,** Comparison of liquid propane and liquid nitrogen in tissue preservation, *Cryobiology,* 2, 342, 1966.
24. **Fishbein, W. N. and Stowell, R. E.,** Mechanism of freezing damage to mouse liver using a mitochondrial enzyme assay. II. Comparison of slow and rapid cooling rates, *Cryobiology,* 6, 227, 1969.
25. **Winckler, J.,** Quenching of tissue in liquid nitrogen-propane, *Histochemie,* 23, 44, 1970.
26. **Ambrosio, L., Inserra, A., and Sfogliano, C.,** Cases of occupational poisoning among workers engaged in bottling commercial liquid gas (butane-propane), *Folia Med.* (Naples), 51, 14, 1968.
27. **Baldock, D. V.,** Attempted murder by calor gas, *Forensic Sci. Soc. J.,* 10, 175, 1970.
28. **Watanabe, S., Kitaguchi, T., Kiyofuji, T., Morisaki, Y., Masuda, T., Noguchi, K., and Matsumoto, S.,** An autopsy case of CO gas poisoning by incomplete combustion of fuel propane gas, *J. Kumamoto Med. Soc.,* 44, 354, 1970.
29. **Okada, H.,** Effects of anoxia before cardiac arrest on postmortem chemical changes in heart muscle, *Shikoku Igaku Zasshi,* 26, 159, 1970.
30. **Inserra, A., Donpe, M., and Timpanaro, V.,** Blood dyscrasia and experimental poisoning with propane and butane, *Folia Med.* (Naples) 51, 779, 1968.
31. **Aviado, D. M. and Belej, M. A.,** Toxicity of aerosol propellants on the respiratory and circulatory systems. I. Cardiac arrhythmia in the mouse, *Toxicology,* 2, 31, 1974.
32. **Belej, M. A., Smith, D. G., and Aviado, D. M.,** Toxicity of aerosol propellants in the respiratory and circulatory systems. IV. Cardiotoxicity in the monkey, *Toxicology,* 2, 381, 1974.
33. **Krantz, J. C., Carr, J., and Vitcha, J. F.,** Anesthesia XXXI. A study of cyclic and noncyclic hydrocarbons on cardiac automaticity, *J. Pharmacol. Exp. Ther.,* 94, 315, 1948.
34. **Aviado, D. M.,** Toxicity of aerosol propellants in the respiratory and circulatory systems. X. Proposed classification, *Toxicology,* 3, 321, 1975.
35. **Aviado, D. M. and Smith, D. G.,** Toxicity of aerosol propellants in the respiratory and circulatory systems. VIII. Respiration and circulation in primates, *Toxicology,* 3, 241, 1975.
36. **Brown, W. E. and Henderson, V. E.,** Experiments with anesthetic gases, *J. Pharmacol. Exp. Ther.,* 27, 1, 1925.

37. **Mehard, C. W., Lyons, J. M., and Kumamoto, J.**, Utilization of model membranes in a test for the mechanism of ethylene action, *J. Membr. Biol.*, 3, 173, 1970.
38. **Mehard, C. W. and Lyons, J. M.**, A lack of specificity for ethylene-induced mitochondrial changes, *Plant Physiol.*, 46, 36, 1970.
39. **Pennington, K. and Fuerst, R.**, Biochemical and morphological effects of various gases on rabbit erythrocytes, *Arch. Environ. Health*, 22, 476, 1971.
40. **Watanabe, K. and Takesue, S.**, Effect of some hydrocarbon gases on egg-white lysozyme activities on different substrates, *Enzymologia*, 41, 99, 1971.
41. **Patty, F. A. and Yant, W. P.**, Odor intensity and symptoms produced by commercial propane, butane, pentane, hexane, and heptane vapor, Bur. Mines Rep. of Investigations No. 2979, 1929.
42. **Kaempfer, H.**, Propan-butan als Treibmittel für Aerosole, *Aerosol Rep.*, 6, 178, 1967.
43. **Chambers, H. E.**, Manufacture and use of butane, *Oil Gas J.*, 50, 67, 1939.
44. **Kiermeier, F. and Stroh, A.**, Untersuchungen und Betrachtungen zur Anwendung von Kunststoffen für Lebensmittel. XV. Mitteilung Über Reaktionen von Polyathylen mit Milch, *Z. Lebensm. Unters. Forsch.*, 14, 208, 1969.
45. **Gordon, R. J., Mayrsohn, H., and Ingels, R. M.**, C_2-C_5 hydrocarbons in the Los Angeles atmosphere, *Environ. Sci. Technol.*, 2, 1117, 1968.
46. **Altshuller, A. P., Lonneman, W. A., Sutterfield, F. D., and Kopczynski, S. L.**, Hydrocarbon composition of the atmosphere of the Los Angeles Basin – 1967, *Environ. Sci. Technol.*, 5, 1009, 1971.
47. **Shugaev, B. B.**, Concentrations of hydrocarbons in tissues as a measure of toxicity, *Arch. Environ. Health*, 18, 878, 1969.
48. **Shugaev, B. B.**, Distribution and toxicity of aliphatic hydrocarbons in body tissues, *Farmakol. Toksikol.* (Moscow), 31, 360, 1968.
49. **Shugaev, B. B.**, Combined action of aliphatic hydrocarbons, such as butane and isobutylene, with reference to their effective concentrations in brain tissue, *Farmakol. Toksikol.* (Moscow), 30, 102, 1967.
50. **Stoughton, R. W. and Lamson, P. D.**, The relative anaesthetic activity of the butanes and pentanes, *J. Pharmacol. Exp. Ther.*, 58, 74, 1936.
51. **Chenoweth, M. B.**, Ventricular fibrillation induced by hydrocarbons and epinephrine, *J. Ind. Hyg. Toxicol.*, 28, 151, 1946.
52. **Heus, J. M. and Glasson, W. A.**, Hydrocarbon reactivity and eye irritation, *Environ. Sci. Technol.*, 2, 1109, 1968.
53. **Altshuller, A. P., Kopczynski, S. L., Wilson, D., Lonneman, W. and Sutterfield, F. D.**, Photochemical reactivities of *n*-butane and other paraffinic hydrocarbons, *J. Air Pollut. Control Assoc.*, 19, 787, 1969.
54. **Frommer, U., Ullrich, V., and Staudinger, H.**, Hydroxylation of aliphatic compounds by liver microsomes. I. Distribution patterns of isomeric alcohols, *Hoppe-Seyler's Z. Physiol. Chem.*, 351, 903, 1970.
55. **Frommer, U., Ullrich, V., and Staudinger, H.**, Hydroxylation of aliphatic compounds by liver microsomes. II. Effect of phenobarbital induction in rats on specific activity and cytochrome P-450 substrate binding spectra, *Hoppe-Seyler's Z. Physiol. Chem.*, 351, 913, 1970.
56. **Mazziotti, A.**, Intermolecular basis of odour, *Nature*, 250, 645, 1974.
57. **Rode, L. J. and Forster, J. W.**, Gaseous alkanes and development in some diverse biological systems, *Z. Allg. Mikrobiol.*, 6, 353, 1966.
58. **Watanabe, K. and Takesu, S.**, Interaction of some gaseous hydrocarbons with egg-white lysozyme, *Hoppe-Seyler's Z. Physiol. Chem.*, 355, 184, 1974.
59. **Ku, H. S. and Leopold, A. C.**, Mitochondrial responses to ethylene and other hydrocarbons, *Plant Physiol.*, 46, 842, 1970.
60. **Battigelli, M. C.**, Air pollution from diesel exhaust, *J. Occup. Med.*, 5, 54, 1963.
61. **Patton, H. W. and Touey, G. P.**, Gas chromatographic determination of some hydrocarbons in cigarette smoke, *Anal. Chem.*, 28, 1685, 1956.
62. **Gleason, M. N., Gosselin, R. E., Hodge, H. C., and Smith, R. P.**, *Clinical Toxicology of Commercial Products, Acute Poisoning*, 3rd ed., Williams & Wilkins, Baltimore, 1969.
63. **Bradbury, F. R. and O'Carroll, F. M.**, Measurement of narcotic potency among houseflies, *Ann. Appl. Biol.*, 57, 15, 1966.
64. **Friedman, S. A., Cammarato, M., and Aviado, D. M.**, Toxicity of aerosol propellants on the respiratory and circulatory systems. II. Respiratory and bronchopulmonary effects in the rat, *Toxicology*, 1, 345, 1973.
65. **Watanabe, K. and Takesue, S.**, Inhibitory effect of some gaseous hydrocarbons on the cell-lysis of *Micrococcus lysodeikticus* by egg-white lysozyme, *Agric. Biol. Chem.*, 36, 825, 1972.
66. **O'Brien, W. E. and Brown, L. R.**, The catabolism of isobutane and other alkanes by a member of the genus Mycobacterium, *Dev. Ind. Microbiol.*, 9, 389, 1967.
67. **Finney, D. J.**, *Probit Analysis*, 3rd ed., Cambridge University Press, 1971.
68. **Aviado, D. M., Zakhari, S., Simaan, J. A., and Ulsamer, A. G.**, *Methyl Chloroform and Trichloroethane in the Environment*, CRC Press, Cleveland, 1976.
69. **Zakhari, S., Leibowitz, M., Levy, P., and Aviado, D. M.**, *Isopropanol and Ketones in the Environment*, CRC Press, Cleveland, 1976.

Part III
Aerosol Formulations

Chapter 7

HYDROCARBON MIXTURE: PROPANE, BUTANE, AND ISOBUTANE

Until recently, fluorocarbons were the major propellants used in the aerosol industry. However, the potential health hazards associated with the use of fluorocarbons and other considerations have made it necessary to develop aerosols using hydrocarbons as propellants. The propellant mixtures consisting of propane, butane, and isobutane described from 1966 to 1970 are becoming important because they represent a means of continuing to use aerosols without fluorocarbons.[1-4]

A new hydrocarbon mixture has been developed specifically for the purpose of substituting for fluorocarbons. The mixture called A-46 has a vapor pressure of 46 psig (at 21.1°C) resulting from the appropriate blending of the following hydrocarbons, having the respective vapor pressures at 21.1°C and the concentrations indicated:

Isobutane (31 psig)	80.4%
Butane (17 psig)	2.5%
Propane (108 psig)	17.1%
Mixture A-46 (46 psig) =	100%

The purpose of this chapter is to describe an investigation to determine whether the mixture of the three hydrocarbons is more toxic than each of the individual components.

A. Inhalational Toxicity in Mice

The LC_{50} of A-46 was determined using male mice (CF-1 strain, Charles River Laboratories, Wilmington, Massachusetts) according to the method described previously (Chapter 2). All gaseous mixtures were balanced by adding oxygen (25% of the volume of A-46), and mortality rate was calculated after a 120-min exposure period.

Results are summarized in Table 7.1. The LC_{50} of A-46 is 57.42% v/v with fiducial limits of 53.96 and 60.88%. There is no significant difference between the LC_{50} of A-46 and that of isobutane alone (52.04% ± 3.26%, as reported in Chapter 6). However, the tendency of A-46 to exhibit less toxicity than isobutane alone might be attributed to the presence of propane which is less toxic than butane or isobutane.

B. Hemodynamic Effects in Dogs

The myocardial and hemodynamic effects of butane, isobutane, and propane in an open-chest dog preparation have been previously reported (Chapters 4, 5, and 6). This is the continuation of an investigation started in this laboratory concerning various gaseous propellants. The hemodynamic effect of a blend of 17.1% propane, 80.4% isobutane, and 2.5% butane (v/v), a hydrocarbon propellant mixture known as A-46, was studied in the same animal preparation along similar lines as methylene chloride (Chapter 2). All data were analyzed by the t-test for paired replicates. The chi-squared test was used to test the association of observed and expected responses. In both cases, the criterion for significance was P less than or equal to 0.05.

The various myocardial and hemodynamic responses to inhalation of various concentrations of A-46 in a typical experiment are illustrated in Figure 7.1. Results are summarized in Table 7.2. The most prominent effect is exhibited on the myocardium. The various effects are as follows.

Decrease in myocardial contractility – Myocardial contractility, as gauged by the maximal rate of rise of left ventricular dp/dt, is attenuated after inhalation of various concentrations of A-46. Thus, inhalation of 0.5%, 1.0%, 2.5%, 5.0%, and 10.0% was accompanied by a decrease in myocardial contractility of 0.6%, 4.6%, 6.3%, 8.5%, and 12.6%, respectively. The negative inotropic effect of A-46 on the heart was not accompanied by a similar negative chronotropic effect; no significant changes were observed in the heart rate.

Decrease in cardiac output – Inhalation of the

TABLE 7.1

Acute Inhalation Toxicity of a Mixture of Propane, *n*-Butane, and Isobutane in Mice after 120 min of Exposure*

Group	Concentration % v/v	% Mortality	LC_{50}	Regression coefficient
1	45	0		
2	50	20	57.42	0.9811
3	55	50	±	
4	60	60	3.46%	
5	70	80		
6	75	100		

*Each group consisted of ten mice.

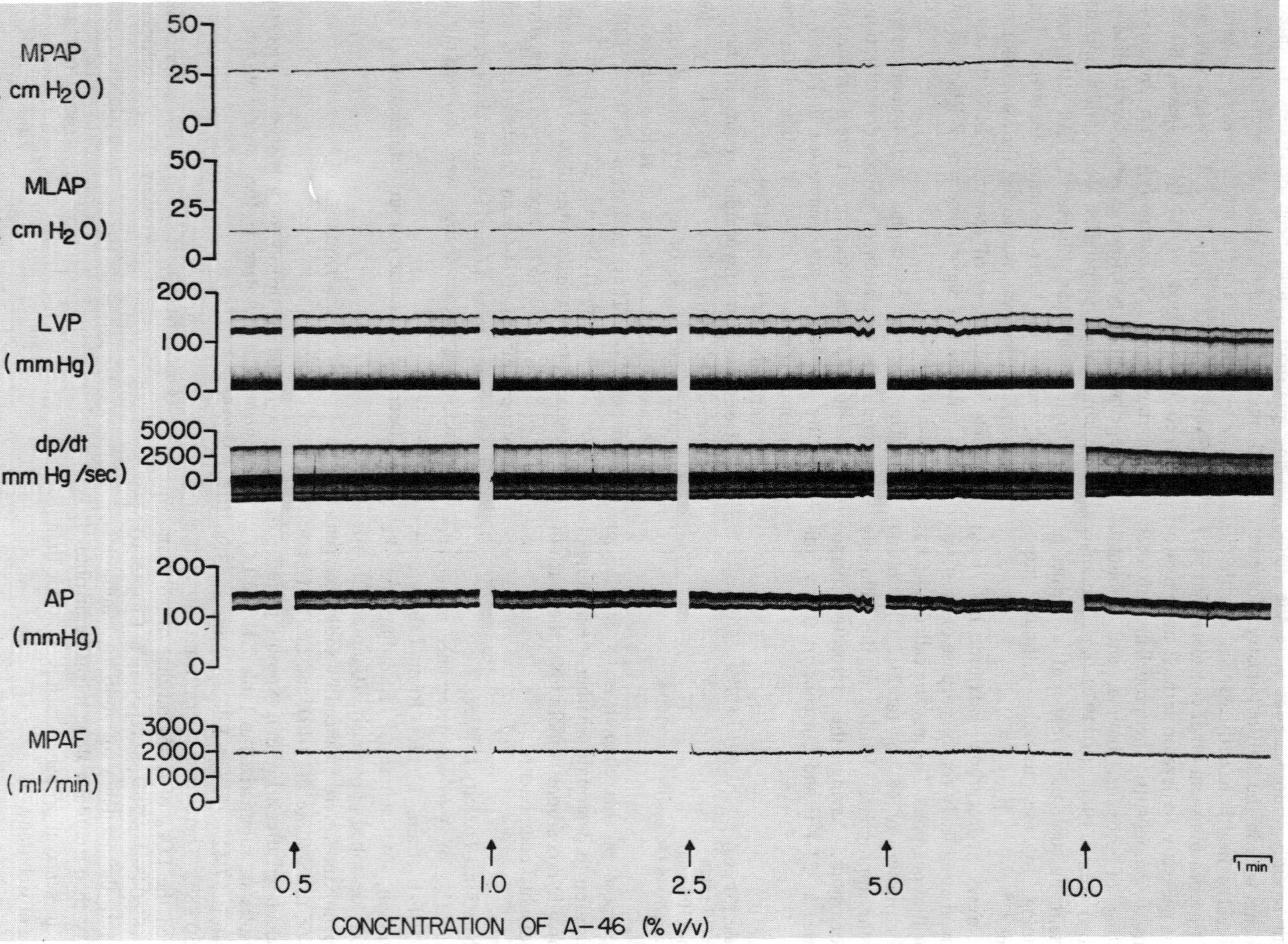

FIGURE 7.1. The effect of progressively increasing concentrations of A-46 on myocardial and hemodynamic parameters in the open-chest dog. MPAP = mean pulmonary arterial pressure; MLAP = mean left atrial pressure; LVP = left ventricular pressure; dp/dt = maximal rate of rise of left ventricular pressure; AP = aortic pressure; MPAF = mean pulmonary arterial flow.

TABLE 7.2

The Myocardial and Hemodynamic Effects of Progressively Increasing Concentrations of A-46 in Dogs*

	MPAP cm H_2O	MLAP cm H_2O	EMPAP cm H_2O	LVP mmHg	LVEDP mmHg	dp/dt mmHg/sec	MAP mmHg	MPAF ml/min	HR beats/min	Vascular resistance dynes·sec/cm^5 Pulmonary	Vascular resistance dynes·sec/cm^5 Systemic	Stroke vol ml	Stroke work g·meter
Control	25.2 ± 0.9	8.0 ± 1.6	17.2 ± 2.2	139 ± 10	1.5 ± 1.0	2945 ± 423	113 ± 7	1880 ± 171	174 ± 10	552 ± 90	4892 ± 523	11.0 ± 1.3	17.7 ± 1.9
A-46 0.5%	25.5 ± 1.3 +0.3 ± 0.2 NS	7.8 ± 1.9 −0.2 ± 0.2 NS	17.7 ± 2.6 +0.5 ± 0.3 NS	138 ± 11 −1 ± 0.6 NS	3.1 ± 2.4 +1.6 ± 2.4 NS	2907 ± 357 −38 ± 54 NS	111 ± 6 −2 ± 1.0 0.05	1842 ± 202 −38 ± 24 NS	174 ± 10 0 ± 0.5 NS	581 ± 103 +29 ± 20 NS	4846 ± 727 −46 ± 72 NS	10.9 ± 1.7 −0.1 ± 0.1 NS	16.9 ± 2.1 −0.8 ± 0.4 NS
A-46 1.0%	25.0 ± 1.1 −0.2 ± 1.3 NS	7.8 ± 1.7 −0.2 ± 0.2 NS	17.2 ± 1.7 0 ± 1.2 NS	136 ± 9 −3 ± 1.2 NS	1.5 ± 1.0 0 ± 0.8 NS	2800 ± 386 −145 ± 48 0.05	109 ± 8 −4 ± 1.3 0.05	1750 ± 196 −130 ± 34 0.02	172 ± 12 −2 ± 2.0 NS	579 ± 90 +27 ± 35 NS	5094 ± 619 +202 ± 127 NS	10.5 ± 1.5 −0.5 ± 0.3 NS	16.0 ± 2.0 −1.7 ± 0.4 0.01
A-46 2.5%	24.3 ± 2.0 −0.9 ± 2.4 NS	7.8 ± 1.9 −0.2 ± 0.3 NS	16.5 ± 2.1 −0.7 ± 2.3 NS	134 ± 10 −5 ± 1.3 0.02	2.0 ± 0.5 +0.5 ± 0.5 NS	2770 ± 411 −175 ± 63 0.05	107 ± 7 −6 ± 1.4 0.02	1760 ± 199 −120 ± 49 0.05	173 ± 12 −1 ± 1.9 NS	571 ± 90 +19 ± 77 NS	5050 ± 620 +158 ± 164 NS	10.3 ± 1.4 −0.7 ± 0.2 0.02	15.6 ± 1.8 −2.1 ± 0.3 0.01
A-46 5.0%	25.3 ± 1.2 +0.1 ± 1.4 NS	7.9 ± 2.0 −0.1 ± 0.5 NS	17.4 ± 1.5 −0.2 ± 1.0 NS	130 ± 9 −9 ± 3.6 0.02	2.6 ± 1.0 +1.1 ± 1.0 NS	2775 ± 430 −170 ± 60 0.05	102 ± 6 −11 ± 3.3 0.02	1730 ± 192 −150 ± 42 0.02	176 ± 13 +2 ± 3.3 NS	623 ± 95 +71 ± 28 NS	4842 ± 543 −50 ± 105 NS	10.0 ± 1.3 −1.0 ± 0.1 0.001	14.3 ± 1.7 −3.4 ± 0.6 0.01
A-46 10.0%	24.1 ± 2.3 −2.1 ± 2.3 NS	7.9 ± 1.6 −0.1 ± 0.7 NS	16.2 ± 1.5 −1.0 ± 1.8 NS	121 ± 7 −18 ± 5 0.02	2.6 ± 0.7 +1.1 ± 0.9 NS	2580 ± 390 −365 ± 114 0.05	96 ± 5 −17 ± 3.3 0.01	1680 ± 192 −200 ± 42 0.01	170 ± 14 −4 ± 5.4 NS	570 ± 95 +18 ± 50 NS	4717 ± 556 −175 ± 142 NS	10.0 ± 1.3 −1.0 ± 0.2 0.02	13.7 ± 1.7 −4.0 ± 0.5 0.01

*Each group of numbers represents the mean value of six experiments, and consists of mean response ± SEM, mean difference ± SE of difference, and the significance level.

previously mentioned concentrations of A-46 was associated with a decrease of 38, 130, 120, 150, and 200 ml/min in mean pulmonary arterial flow. The decrease in cardiac output was almost parallel to the decrease in myocardial contractility (Figure 7.2). Since there was no change in heart rate, the decrease in cardiac output was therefore accompanied by a decrease in stroke volume.

Decrease in left ventricular pressure and mean aortic pressure – The decrease in left ventricular pressure following the inhalation of 0.5%, 1.0%, 2.5%, 5.0%, and 10.0% of A-46 averaged to 0.7%, 2.2%, 3.4%, 6.4%, and 12.8% respectively. The mean corresponding decreases in aortic pressure were 2.3%, 3.9%, 5.3%, 10.0% and 14.7%, respectively. The combined effect of A-46 on the heart and blood vessels was shown by the decrease in stroke work of 4.2%, 9.8%, 11.9%, 18.8%, and 22.6%, after inhalation of 0.5%, 1.0%, 2.5%, 5.0%, and 10%, respectively. No remarkable changes were observed in pulmonary arterial pressure or left atrial pressure.

Comparison of propane, butane, isobutane, and the hydrocarbon mixture – The effects of propane, butane, and isobutane have been previously studied in this laboratory in a similar open-chest dog preparation. The most prominent effect, in fact, an effect that is observed with the lowest concentration of all the studied hydrocarbons, is myocardial depression. Figures 7.3 and 7.4 show the comparative myocardial and hemodynamic effects of all these gases. The differences among the effects of these compounds seem to be quantitative and are summarized in Tables 7.3 and 7.4. These three hydrocarbons can be arranged, according to our findings, in the following order of descending toxicity:

butane > isobutane = A-46 > propane

C. Discussion of Hemodynamic Effects

Results of the present investigation show that A-46, like other saturated gaseous aliphatic hydrocarbons, possesses a myocardial depressant effect. This effect is shown by the decrease in myocardial contractile force as gauged by the changes in the left ventricular dp/dt. Since the decrease in cardiac output is almost parallel to the decrease in myocardial contractile force, it is most probable that the decrease in the former parameter is due mainly to the decrease in the latter. However, the possibility of a decrease in venous return, with the consequent decrease in ventricular filling pressure as a contributing mechanism for the decrease in cardiac output, cannot be excluded.

The decrease in systemic arterial pressure is an expected consequence of attenuation of myocardial contraction and cardiac output. The depressant effect of these gases on the central nervous system with the consequent depression of vasomotor tone might contribute to the observed decrease in the systemic blood pressure.

Comparison of the effects of various gaseous

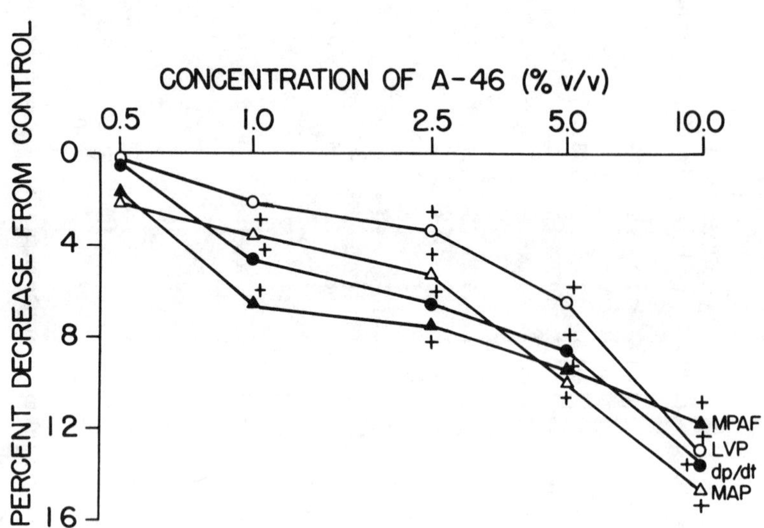

FIGURE 7.2. Mean percentage changes from control averages in response to progressively increasing concentrations of A-46. Abbreviations are the same as for Figure 7.1. Asterisk denotes significant changes. Bars representing standard errors of means are omitted for simplicity.

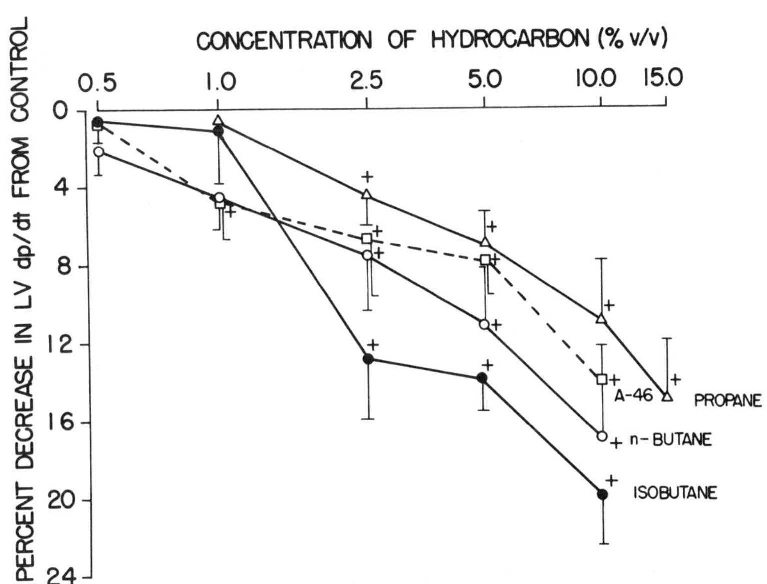

FIGURE 7.3. Concentration-response curves of the myocardial depressant effects of various aliphatic hydrocarbons in the open-chest dog preparation. Note that propane is the least toxic of all the studied compounds.

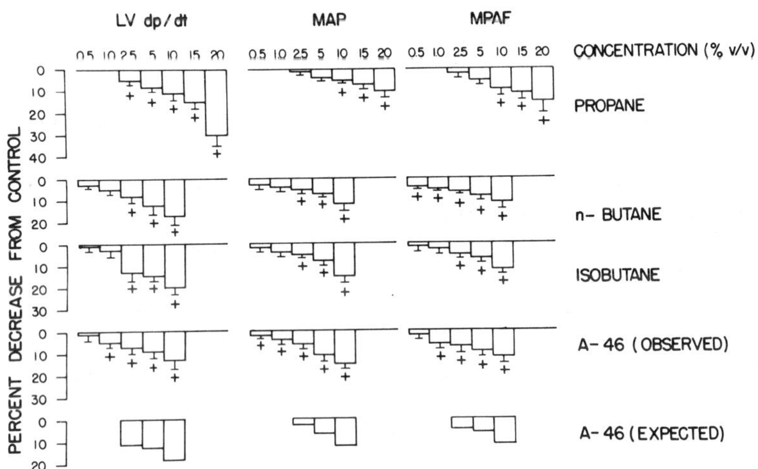

FIGURE 7.4. Mean percent decrease in myocardial contractility (dp/dt), mean aortic pressure (MAP), and mean pulmonary arterial flow (MPAF), brought about by various hydrocarbons studied. The expected effects of A-46 were calculated according to the molar percent of each component.

TABLE 7.3

Comparison of Propane, *n*-Butane, Isobutane, and A-46

	Propane	*n*-Butane	Isobutane	A-46
Formula	$CH_3-CH_2-CH_3$	$CH_3-CH_2-CH_2-CH_3$	$CH_3-CH(CH_3)-CH_3$	19.7 mol % propane 77.8 mol % isobutane 2.5 mol % *n*-butane
Molecular weight	44.09	58.12	58.12	
Boiling point	−42.1°C	−0.5°C	−0.5°C	
Solubility	All are soluble in alcohol, ether, chloroform			
Threshold effective concentration*	3.3%	0.5%	2.0%	1.9%

*Values as obtained in this laboratory.

hydrocarbons studied in this laboratory shows that propane is the least toxic, having a threshold effective concentration (TEC) of 3.3%; butane is the most toxic (TEC 0.5%), and isobutane and A-46 are intermediate and almost identical (TEC 2.0% and 1.9%, respectively). At higher concentrations, e. g., 5.0% and 10.0%, butane and isobutane exert comparative hemodynamic effects. The combination of 19.7 mol % of propane, 77.8 mol % of isobutane, and 2.5 mol % of *n*-butane to give A-46 seems to show an inclination to a decrease in myocardial toxicity, though not significantly different from that of isobutane alone. The expected effects of A-46, calculated according to the molar concentration of various components versus the observed values, are depicted in Figure 7.4. There is no significant difference between the expected and observed values which indicates that there is no potentiation of myocardial depression brought about by the combination of gases at the concentration levels studied. Propane, on the other hand, seems to possess the lowest toxicity of all studied compounds. The basis of selection of a suitable propellant from these gases depends, among other factors, on various physicochemical properties, stability, volatility, vapor pressure, etc., which are beyond the scope of this report.

D. Summary

The inhalation of the hydrocarbon propellant mixture A-46 in the anesthetized open-chest dog preparation brought about the following hemodynamic changes: a) decrease in myocardial contractility, b) decrease in cardiac output, and c) decrease in left ventricular pressure and systemic blood pressure. The minimum effective concentration was 1.9 ± 0.4% v/v. The propellant A-46 possesses the same pharmacologic effect as other studied hydrocarbons, though it has a tendency to be less toxic than isobutane alone. With the hydrocarbon propellant mixtures, there is no potentiation of the individual effects of isobutane, butane, and propane.

TABLE 7.4

Summary of Percent Changes in Various Parameters by Various Concentrations of Hydrocarbons in Dogs

Concentration		Myocardial contractility				Mean aortic pressure				Mean pulmonary arterial flow				Pulmonary vascular resistance				Systemic vascular resistance				Stroke work			
% v/v	m*M*/l	P	*n*-B	i-B	A-46	P	*n*-B	i-B	A-46	P	*n*-B	i-B	A-46	P	*n*-B	i-B	A-46	P	*n*-B	i-B	A-46	P	*n*-B	i-B	A-46
0.5	0.20	–	-2.3	-0.5	-0.6	–	-2.8	-2.8	-2.3*	–	-4.1*	-1.9	-2.0	–	+7.9	+2.2	+4.9	–	-0.2	-1.3	-1.6	–	-5.9*	-4.8	-4.2
1.0	0.41	–	-4.5	-0.6	-4.6*	–	-3.0	-0.4	-3.9*	–	-4.8*	-2.1	-6.9*	–	+10.7	+1.1	+10.0	–	+1.1	-2.0	+3.3	–	-7.8*	-2.1	-9.8*
2.5	1.02	-4.5*	-7.5*	-13.1*	-6.3*	-0.1	-3.8*	-3.7*	-5.3*	-1.3	-6.0*	-5.3*	-7.0*	+2.6	+10.4	-0.4	+7.1	+1.6	+0.8	-0.3	+2.5	-2.6	-9.2*	-9.8*	-11.9*
5.0	2.04	-7.2*	-11.2*	-13.8*	-8.5*	-3.1*	-5.7*	-7.8*	-10.0*	-4.7	-7.4*	-6.2*	-8.5*	+7.2*	+12.4	+4.9	+13.9	+1.9	-0.3	+4.1	-1.2	-10.3*	-13.6*	-15.4*	-18.8*
10.0	4.09	-11.0*	-17.1*	-19.8*	-12.6*	-4.3*	-12.4*	-15.0*	-14.7*	-8.4*	-10.1*	-12.3*	-11.2*	+10.3*	+11.1	+16.5*	+5.7	+5.0	-5.5	-3.2	-4.0	-15.0*	-21.7*	-25.6*	-22.6*
15.0	6.13	-14.9*	–	–	–	-6.9*	–	–	–	-10.8*	–	–	–	+11.1*	–	–	–	+3.3	–	–	–	-18.0*	–	–	–
20.0	8.18	-29.4*	–	–	–	-9.3*	–	–	–	-14.7*	–	–	–	+17.2*	–	–	–	+2.2	–	–	–	-24.2*	–	–	–

Note: The various concentrations studied are expressed in m*M*/l and % v/v. P = propane, *n*-B = *n*-butane, *i*-B = isobutane. A-46 consists of the following molar percentages: 19.7 propane, 77.8 isobutane, 2.5 *n*-butane. See Section C, Discussion of Hemodynamic Effects.

*Significant change ($P < 0.05$).

Chapter 8

INTERACTIONS AMONG HYDROCARBON PROPELLANTS, METHYLENE CHLORIDE, AND ETHANOL

Although several aerosol formulations are in current use, there has been no previous attempt to investigate the interaction of the constituents. This chapter is the first reported attempt to examine the interaction of a new formulation containing the following ingredients discussed elsewhere in this monograph: methylene chloride (Chapter 2); ethanol (Chapter 3); propane (Chapter 4); butane (Chapter 5); and isobutane (Chapter 6). The last three are hydrocarbon propellants contained in mixture A-46 (Chapter 7).

A. Methods and Calculations

Experiments were carried out on adult mongrel dogs of either sex weighing between 13 and 24 kg (average 18 kg), anesthetized with intravenous injection of 30 to 35 mg/kg of pentobarbital sodium. After instituting artificial respiration, arrangements were made to measure cardiac output, myocardial force of contraction, left atrial pressure, left ventricular pressure, pulmonary arterial pressure and heart rate, as previously described (Chapter 2).

After allowing the preparation to stabilize for 30 min the threshold effective concentrations of A-46 (17.1% propane, 80.4% isobutane, and 2.5% butane v/v), ethanol, methylene chloride, and various combinations of these compounds, were administered via the inlet of the respirator for 5 min. The preparation was allowed to recover before administration of each subsequent compound. The sequence of inhalation was altered from one preparation to the next. All data were analyzed by paired comparison. The Student t-test was used to test the significance of difference between various mixtures. In both cases, the criterion for significance is P less than or equal to 0.05. The calculations of hemodynamic parameters and abbreviations are described in Chapter 2.

B. Results

Interaction between A-46 and methylene chloride – The mixture of gaseous propellants known as A-46 induced various hemodynamic changes when inhaled in a 2% v/v concentration in the open-chest dog preparation (Table 8.1). These changes comprised a decrease in left ventricular pressure, myocardial contractility as indicated by maximal rate of rise of left ventricular pressure (dp/dt), mean arterial pressure, systemic vascular resistance, stroke volume, and stroke work averaging to 2.4%, 4.5%, 2.9%, 1.1%, 5.6%, and 8.8%, respectively. An increase of 3.1% in the heart rate was also induced by this concentration. Nonsignificant changes in other measured or derived parameters were observed.

On the other hand, inhalation of 1% v/v concentration of methylene chloride, in the same preparation, brought about a slight but significant increase in mean arterial pressure (2.1%) and stroke work (2.3%).

Inhalation of a mixture of 2% A-46 and 1% methylene chloride brought about an increase in systemic vascular resistance of 6.2% and a decrease in stroke volume of 3.9%. The hemodynamic and myocardial effects of A-46 and/or methylene chloride are depicted in Figure 8.1. The myocardial effects of A-46 predominate over those of methylene chloride, while the vascular effects of the latter have the upper hand when mixed with the former. Thus, while methylene chloride brings about a slight increase in left ventricular pressure (0.4%) and cardiac output (0.2%), A-46 decreases the same parameters by 2.4% and 2.2%, respectively. The effect of inhaling a mixture of both, however, revealed a decrease of 1.1% in the left ventricular pressure and 2.6% in cardiac output, as compared with the average control. The predominance of the effect of A-46 on the heart manifests itself in a decrease in stroke volume. Thus, while methylene chloride alone brought about a decrease in stroke volume averaging 0.4%, and A-46 decreased the same parameter by 5.6%, the effect of the mixture of A-46 and methylene chloride is a decrease of 3.9%, which is significantly different from the effect of methylene chloride alone.

On the other hand, the increase in mean arterial pressure (2.1%) induced by methylene chloride alone not only annuls the decrease in the same parameter (2.9%) under the influence of A-46 alone, when both are administered together, but also brings about a net increase of 2.9% which is significantly different from that of A-46 alone. The same is true in the case of stroke work. Thus, while A-46 alone brought about a decrease in this

TABLE 8.1

The Effect of Inhalation of 2% A-46, 1% Methylene Chloride, and a Combination of Both on the Hemodynamics of the Open-chest Dog

	A-46			A-46 + Meth. Chloride			Methylene Chloride		
		Response			Response			Response	
	Control		Δ %	Control		Δ %	Control		Δ %
MPAP cm H_2O	25.5 ±2.9	26.2 ±2.9 NS	+2.8 ±2.6	25.9 ±3.2	26.4 ±2.9 NS	+2.6 ±2.3	26.7 ±2.6	27.8 ±3.1 NS	+3.5 ±2.9
MLAP cm H_2O	7.9 ±0.9	7.5 ±1.0 NS	−4.8 ±6.9	7.4 ±1.1	7.5 ±0.9 NS	+3.5 ±5.8	7.5 ±0.9	7.8 ±0.9 NS	+4.2 ±3.1
EMPAP cm H_2O	17.6 ±3.4	18.7 ±3.5 NS	+6.7 ±5.8	18.4 ±3.5	18.8 ±3.2 NS	+4.0 ±4.3	19.2 ±2.9	19.9 ±3.7 NS	+2.1 ±4.6
LVP mmHg	136 ±5.1	133 ±5.7 0.02	−2.4 ±0.7	134 ±5.7	132 ±4.7 NS	−1.1 ±1.3	133 ±5.8	134 ±5.7 NS	+0.4 ±1.1
LVEDP mmHg	3.8 ±0.6	4.6 ±0.4 NS	+33.3 ±21.1	5.0 ±1.2	5.4 ±1.1 NS	+5.6 ±5.5	5.4 ±1.0	5.8 ±0.8 NS	+16.7 ±16.6
dp/dt mmHg/sec	2896 ±152	2771 ±169 0.02	−4.5 ±1.3	2854 ±178	2813 ±182 NS	−1.2 ±3.2	2813 ±157	2771 ±200 NS	−1.9 ±2.3
MAP mmHg	112 ±5.7	109 ±5.4 0.05	−2.9 ±1.1	109 ±5.6	112 ±6.0 NS	+2.9 ±1.6	113 ±6.2	115 ±6.2 0.001	+2.1 ±0.2
MPAF ml/min	1808 ±241	1775 ±250 NS	−2.2 ±1.1	1783 ±251	1733 ±240 NS	−2.6 ±1.3	1767 ±273	1758 ±253 NS	+0.2 ±1.3
HR beats/min	179 ±8.6	184 ±7.9 0.05	+3.1 ±1.4	184 ±7.9	186 ±7.2 NS	+1.2 ±0.7	186 ±7.1	186 ±7.2 NS	−0.1 ±0.9
PVR dynes·sec/cm^5	614 ±146	663 ±154 NS	+9.8 ±6.3	653 ±157	698 ±164 NS	+6.7 ±3.6	698 ±153	714 ±169 NS	+2.0 ±4.6
SVR dynes·sec/cm^5	5039 ±496	4991 ±500 0.05	−1.1 ±0.4	4901 ±463	5222 ±523 0.01	+6.2 ±1.1	5225 ±579	5325 ±583 NS	+2.2 ±1.1
Stroke vol ml	10.1 ±1.1	9.6 ±1.2 0.01	−5.6 ±1.0	9.7 ±1.2	9.3 ±1.1 0.02	−3.9 ±1.2	9.5 ±1.3	9.4 ±1.3 NS	−0.4 ±0.9
Stroke work g·meter	16.2 ±2.2	15.0 ±2.4 0.01	−8.8 ±1.6	15.3 ±2.5	15.1 ±2.3 NS	−0.9 ±2.7	15.4 ±2.6	15.7 ±2.6 0.05	+2.3 ±0.9

*Each set of numbers is based on six dogs, and consists of mean control ± SEM, mean experimental ± SEM, percent change ± SEM and the significance level. NS = nonsignificant. Abbreviations are explained in the text.

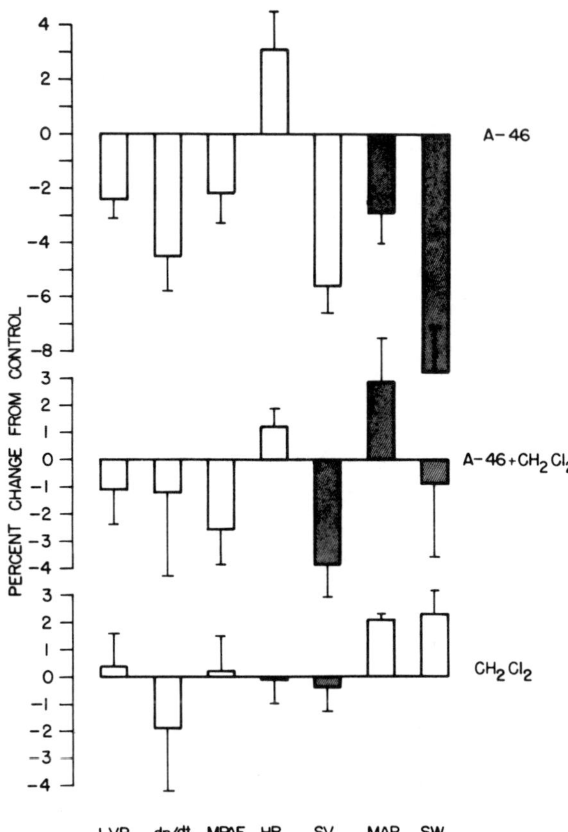

FIGURE 8.1. Percent changes in various parameters after the inhalation of 2% A-46, 1% methylene chloride (CH_2Cl_2), and a mixture of both. LVP = left ventricular pressure; dp/dt = maximum rate of rise of ventricular pressure; MPAF = mean pulmonary arterial flow; HR = heart rate, SV = stroke volume; MAP = mean arterial pressure; SW = stroke work. Hatched bars within the same parameter indicate a significant difference.

parameter of 8.8%, the mixture of A-46 and methylene chloride decreased it by only 0.9%.

Interaction between A-46 and ethanol – In this series of experiments, inhalation of 2.0% v/v of A-46 brought about changes similar to those reported above. Thus, decreases of 2.8%, 5.3%, 4.3%, 6.0%, 7.5%, and 11.6% were induced in left ventricular pressure, left ventricular dp/dt, mean arterial pressure, cardiac output, stroke volume, and stroke work, respectively. Inhalation of ethanol alone (2.5%) brought about a significant decrease in cardiac output, stroke volume, and stroke work, averaging 5.3%, 4.1%, and 5.3%, respectively (Table 8.2.) A significant decrease in effective mean pulmonary arterial pressure averaging to 5.9% was also observed under the influence of 2.5% ethanol. Inhalation of a mixture of 2% A-46 and 2.5% ethanol brought about no significant change in any of the studied parameters, however. Although both A-46 and ethanol share the property of depressing the myocardium when administered separately, neither synergistic nor additive effect has been observed when these compounds are administered together. However, some antagonistic effect has been observed. For instance, a decrease in left ventricular pressure of 2.8% accompanied the administration of A-46 or ethanol separately, while a decrease of only 2% was obtained after inhalation of a mixture of both. The same is valid for other parameters, e. g., left ventricular dp/dt, mean pulmonary arterial flow, stroke volume, and stroke work. The most dramatic change, however, was in mean aortic pressure. Thus, while A-46 and ethanol brought about a decrease in mean aortic pressure of 4.3% and 1.2% respectively, the mixture of both brought about a slight and nonsignificant increase in the same parameter averaging 0.2% (Figure 8.2).

Interaction between ethanol and methylene chloride – Inhalation of 1% methylene chloride brought about an increase in left ventricular pressure, mean arterial pressure, and systemic vascular resistance averaging 2.8%, 2.0%, and 3.0%, respectively (Table 8.2). Nonsignificant decreases in myocardial contractility, stroke volume, and cardiac output were also induced. However, inhalation of a mixture of 2.5% ethanol and 1% methylene chloride brought about, except for a slight increase in mean arterial pressure (1.2%), decreases in left ventricular pressure, maximal rate of rise of left ventricular pressure (dp/dt), cardiac output, stroke volume and stroke work, averaging to 0.3%, 3.6%, 3.5%, 6.1%, and 4.4%, respectively.

Arrhythmogenic effect of methylene chloride and its mixture with ethanol – Inhalation of 5% methylene chloride alone brought about a second degree block in the open-chest dog preparation. Ventricular extrasystole was also observed. Inhalation of a mixture of 2.5% ethanol and 5% methylene chloride did not abolish arrhythmia induced by the latter. However, increasing the concentration to 5%, 7.5%, and 10.0% completely abolished the arrhythmogenic effect of methylene chloride (Figures 8.3 and 8.4).

C. Discussion

On the basis of extensive investigations carried out in this laboratory, the most widely used fluorinated propellants, especially trichlorofluoro-

TABLE 8.2

The Effect of Inhalation of 2% A-46, 2.5% Ethanol, 1% Methylene Chloride, and Various Combinations on the Hemodynamics of Open-chest Dog Preparation

	MPAP cm H_2O		MLAP cm H_2O		EMPAP cm H_2O		LVP mmHg		LVEDP mmHg		dp/dt mmHg/sec		MAP mmHg		MPAF ml/min		HR beats/min		Vascular resistance dynes·sec/cm^5				Stroke vol ml		Stroke work g·meter	
																			Pulmonary		Systemic					
	C	E	C	E	C	E	C	E	C	E	C	E	C	E	C	E	C	E	C	E	C	E	C	E	C	E
2% A-46	24.3	25.2	8.1	7.8	16.2	17.4	132	128	5.4	7.1	2771	2625	107	102	1825	1725	183	186	609	641	4781	4817	10.0	9.3	15.4	13.6
	+0.9 ± 0.4		−0.2 ± 0.2		+1.2 ± 0.5		−4 ± 1.5		+1.7 ± 0.5		−146 ± 38		−5 ± 0.7		−100 ± 34		+3 ± 1.3		+32 ± 39		+36 ± 151		−0.7 ± 0.2		−1.7 ± 0.4	
	+4.4 ± 1.8%		−3.3 ± 3.1%		+9.7 ± 4.1%		−2.8 ± 1.1%		+36.1 ± 15.2%		−5.3 ± 1.6%		−4.3 ± 0.7%		−6 ± 2.4%		+1.5 ± 0.7%		+1.8 ± 5.4%		+0.5 ± 2.6%		−7.5 ± 2.2%		−11.6 ± 2.1%	
	NS		NS		NS		0.05		0.05		0.02		0.01		0.05		NS		NS		NS		0.02		0.01	
2% A-46	22.9	23.0	8.4	8.0	14.5	15.0	130	128	6.0	7.0	2725	2575	106	106	1810	1760	177	177	537	573	4767	5058	10.0	9.8	15.2	14.9
+ 2.5%	+0.1 ± 0.9		−0.4 ± 0.3		+0.5 ± 1.0		−2 ± 1.1		+1 ± 1.0		−150 ± 73		0 ± 0.9		−50 ± 27		−0.6 ± 0.8		+36 ± 30		+291 ± 179		−0.2 ± 0.2		−0.3 ± 0.4	
ethanol	+0.8 ± 4.4%		−5.4 ± 3.6%		+6.3 ± 7.9%		−2.1 ± 1.0%		+20 ± 20%		−4.8 ± 2.4%		+0.2 ± 0.8%		−3.4 ± 2.2%		−0.3 ± 0.4%		+10.1 ± 7.9%		+6.6 ± 3.6%		−3.2 ± 2.4%		−2.6 ± 2.9%	
	NS		NS		NS		NS		NS		NS		NS		NS		NS		NS		NS		NS		NS	
2.5%	25.5	24.5	7.8	7.9	17.6	16.6	128	125	5.4	7.5	2646	2521	103	102	1800	1700	184	184	674	673	4973	5083	9.2	8.8	13.8	12.9
ethanol	−1.0 ± 0.5		+0.1 ± 0.3		−1.1 ± 0.4		−3 ± 2.2		+2.1 ± 0.4		−125 ± 56		−1 ± 1.8		−100 ± 31		−0.3 ± 2.5		−0.2 ± 12		+110 ± 190		−0.4 ± 0.1		−0.9 ± 0.3	
	−3.4 ± 1.9%		+0.9 ± 2.7		−5.9 ± 2.3%		−2.8 ± 1.8%		+61.1 ± 18.1%		−4.9 ± 2.1%		−1.2 ± 1.8%		−5.3 ± 1.6%		−0.2 ± 1.4%		−1.6 ± 2.0%		+1.1 ± 2.9%		−4.1 ± 2.2%		−5.3 ± 1.9%	
	NS		NS		0.05		NS		0.01		NS		NS		0.05		NS		NS		NS		0.05		0.05	
2.5%	25.8	25.8	7.5	7.5	18.3	18.3	131	130	6.5	7.0	2675	2575	103	104	1740	1670	183	188	753	755	5199	5488	9.9	9.3	13.2	12.2
ethanol +	0 ± 0.8		0 ± 0.2		0 ± 0.9		−1 ± 1.7		+0.5 ± 0.5		−100 ± 61		+1 ± 2.8		−70 ± 64		+4.2 ± 0.8		+2 ± 28		+289 ± 183		−0.6 ± 0.3		−0.9 ± 0.9	
1% methylene	+0.1 ± 3.6%		+0.2 ± 2.5%		+0.2 ± 5.9%		−0.3 ± 1.3%		+20 ± 20%		−3.6 ± 2.2%		+1.2 ± 2.9		−3.5 ± 3.1%		+2.3 ± 0.5%		+4.5 ± 8.1%		+4.5 ± 2.2%		−6.1 ± 2.9%		−4.4 ± 5.4%	
chloride	NS		NS		NS		NS		NS		NS		NS		NS		0.01		NS		NS		NS		NS	
1% methylene	25.1	25.4	7.4	7.6	17.8	17.8	129	132	6.5	7.0	2600	2600	101	103	1790	1780	187	186	744	739	5361	5609	9.4	9.4	12.0	12.0
chloride	+0.3 ± 0.3		+0.3 ± 0.3		0 ± 0.3		+3 ± 1.9		+0.5 ± 0.5		0 ± 40		+2 ± 0.5		−10 ± 29		−1.0 ± 0.9		−5 ± 25		+248 ± 223		0 ± 0.2		0 ± 0.1	
	+1.1 ± 1.3%		+5.2 ± 5.2%		+0.9 ± 2.2%		+2.8 ± 2.1%		+6.7 ± 6.7%		−0.5 ± 1.7%		+2 ± 1.4%		−1.4 ± 1.6%		−0.6 ± 0.5%		−0.7 ± 2.5%		+3.0 ± 3.1%		−1.5 ± 1.8%		−0.5 ± 0.8%	
	NS		NS		NS		NS		NS		NS		NS		0.05		NS		NS		NS		NS		NS	

*Each set of numbers is based on six dogs, and consists of mean control (C), mean experimental (E), difference ± SE of difference, Δ% ± SE, and the significance level. NS = nonsignificant. Abbreviations are explained in the text.

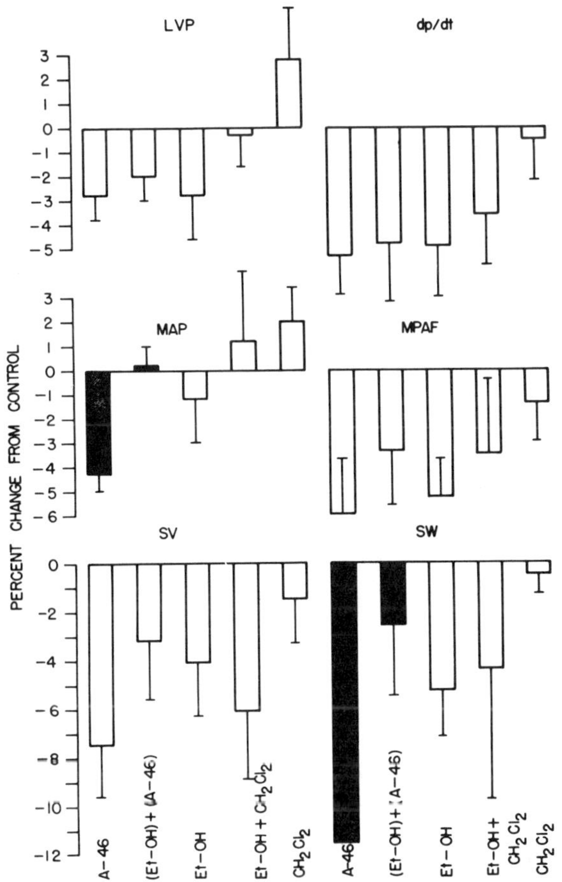

FIGURE 8.2. Percent change from control in various parameters after inhalation of 2% A-46, 2.5% ethanol (Et-OH), and/or 1% methylene chloride (CH_2Cl_2). Abbreviations are the same as for Figure 8.1. There is a significant difference between each two neighboring bars.

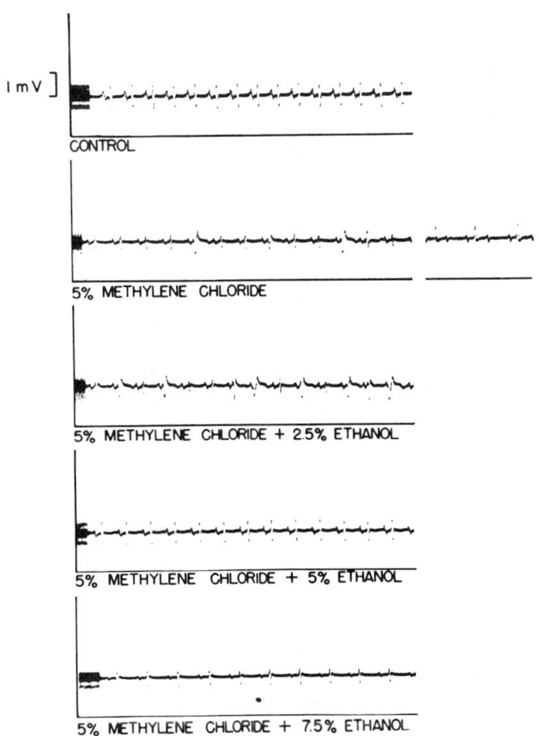

FIGURE 8.3. The electrocardiogram (lead III) before and after inhalation of 5% methylene chloride, alone or in combination with various concentrations of ethanol. Note the heart block developed after methylene chloride. It is abolished after high concentrations of ethanol, which bring about bradycardia as well.

methane (FC 11), were found to be the most toxic in various animal species.[5] Fluorinated hydrocarbons were reported to be especially hazardous to the heart and lungs of various animal species[6] as well as to humans.[7] It was found necessary, therefore, to study a non-fluorinated aerosol formulation that not only performs the same functions as fluorinated propellants, but also proves to be less toxic than the latter. Various halogenated and non-halogenated compounds have been tested (Chapters 2 to 7) and it is the purpose of the present study to discover any interaction between the various components of a non-fluorinated aerosol propellant. For this reason, the threshold effective concentrations of the hydrocarbon propellant mixture known as A-46 (consisting of 17.1% propane, 80.4% isobutane, and 2.5% butane v/v), ethanol, and methylene chloride were tested for any possible interaction in the anesthetized open-chest dog preparation. Results of this study showed that the common feature of all these compounds was that they caused depression of the myocardium as manifested by the decrease in contractility and output of the heart (Figure 8.2). Except for a slight increase in mean arterial and left ventricular pressures induced by methylene chloride, all tested compounds brought about a decrease in left ventricular pressure, mean arterial pressure, stroke volume, and stroke work. The difference in effect seems to be quantitative. Inhalation of 2% of A-46 brought about the most significant changes in various parameters, especially in myocardial contractility and output. Methylene chloride, on the other hand, inhaled in a 1% concentration, brought about a slight increase in cardiac output. However, the main difference between the two compounds, viz., A-46 and methylene chloride, seems to be their effect on the vasculature. Thus, while A-46 decreased systemic vascular resistance by 1.1%, methylene

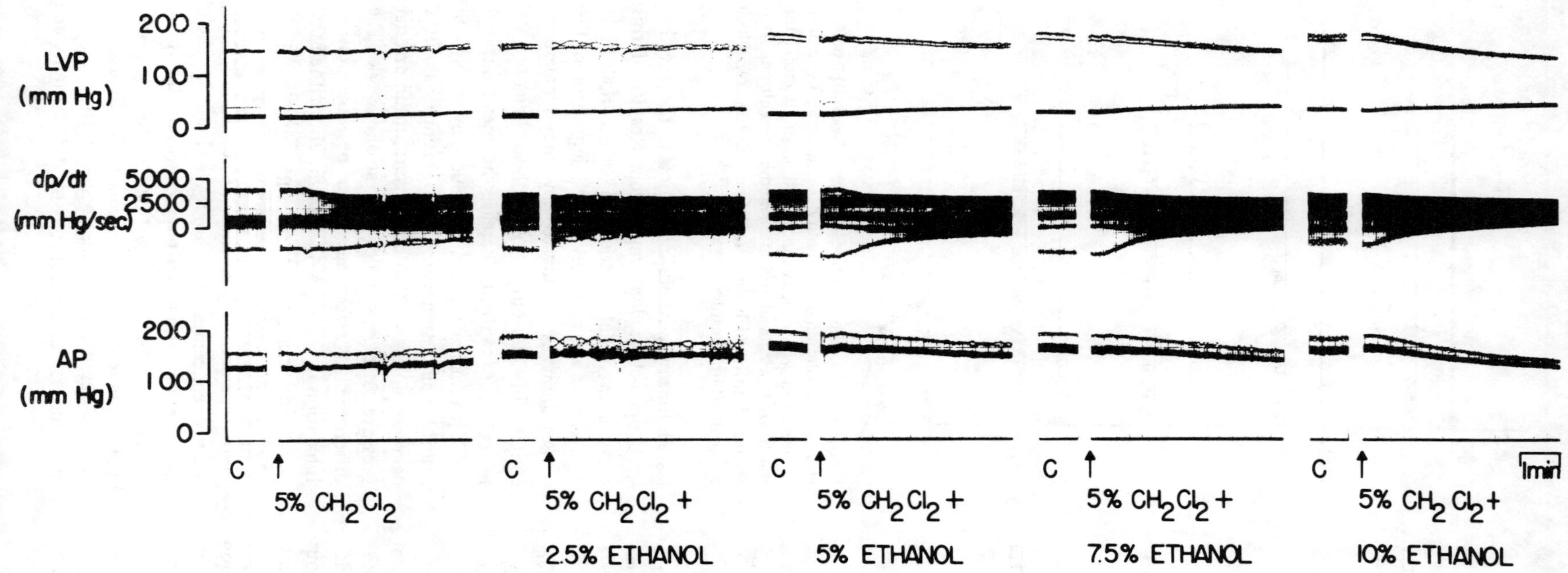

FIGURE 8.4. The effect of methylene chloride alone, or in combination with various concentrations of ethanol, on the open-chest dog preparation. Note arrhythmia induced by methylene chloride alone, which is abolished by simultaneous administration of ethanol, 5, 7.5, or 10% (but not 2.5%).

chloride increased the same parameter by 2.2%. This is also reflected in the mean arterial pressure which, despite the decrease in myocardial contractility, was increased by 2.1% under the influence of methylene chloride. Inhalation of a mixture of both compounds, however, showed that while the effects of A-46 predominate over those of methylene chloride on the myocardium, the opposite is true on the blood vessels. In other words, the net result of the simultaneous inhalation of both compounds is a decrease in cardiac output and stroke volume, the latter being significantly different from the effects of inhalation of methylene chloride alone, and an increase in mean aortic pressure which predominates over the decrease in aortic pressure induced by A-46 alone (see Figure 8.1).

Inhalation of ethanol alone in 2.5% concentration brought about a decrease in several hemodynamic parameters studied. Generally, these decreases are less in magnitude than those elicited by A-46 alone. Nonetheless, simultaneous inhalation of these two myocardial depressant agents, viz., ethanol and A-46, save for a slight increase in mean aortic pressure, resulted in an effect that is qualitatively similar to the effect of both compounds administered separately, yet lesser in magnitude than either. The most striking effect, however, is on aortic pressure. Thus, while A-46 and ethanol brought about a decrease in mean aortic pressure of 4.3% and 1.2%, respectively, a mixture of both elicited a slight increase of 0.2% in the same parameter which is significantly different from the effect of A-46 alone. Since aortic pressure is determined by several factors related to the pumping activity of the heart as well as the tone in various vascular beds, an explanation of this effect would be mere speculation. However, examination of the effect on myocardial contractility and output may shed some light on the subject. Since inhalation of the mixture of A-46 and ethanol together brought about a decrease in the rate of rise of left ventricular pressure (dp/dt) and cardiac output that was lesser in magnitude than the effect of either of these compounds alone, the decrease in mean aortic pressure by the mixture would be expected to be less conspicuous. Furthermore, the effect of ethanol and A-46 on the smooth muscle of various vascular beds, and the possibility of the release of catecholamines from various storage sites, cannot be excluded as contributing factors. It is well documented that acetaldehyde, the metabolic product of ethanol, releases catecholamines from the adrenal medulla.[8,9]

Since the components of A-46 (propane, butane, and isobutane) were reported to affect various enzymatic systems,[10-12] it is possible that they affect alcohol dehydrogenase which metabolizes alcohol to acetaldehyde with the consequent release of catecholamines.

Administration of ethanol and methylene chloride together resulted in myocardial and hemodynamic effects that are not significantly different from either ethanol or methylene chloride alone.

Simultaneous inhalation of 5%, 7.5%, or 10% ethanol and 5% methylene chloride showed that, in these concentrations, ethanol abolished the myocardial arrhythmia brought about by methylene chloride. That ethanol combats the arrhythmogenic effect of methylene chloride is clearly shown in Figures 8.3 and 8.4. This effect might be due to the myocardial depressant effect of ethanol, its effect on the electrolyte pattern of the heart, or its local anesthetic effect.

D. Summary

The myocardial and hemodynamic effects of A-46, methylene chloride, and ethanol were studied in the open-chest dog preparation. Administered separately or in various combinations, these compounds attenuate myocardial contractile force as well as cardiac output. The decrease in cardiac output under the influence of A-46 overshadows that of methylene chloride when both are administered simultaneously. However, the vascular effects of methylene chloride predominate over those of A-46. Some degree of antagonism on systemic blood pressure between A-46 and ethanol was observed. Furthermore, ethanol administered in concentrations of 5%, 7.5%, and 10% combatted the arrhythmogenic effect of methylene chloride.

Chapter 9

CONCLUDING REMARKS

For the first time in this Solvents in the Environment Series, a monograph is being devoted to a group of consumer products, specifically the solvents and propellants used in aerosols. During the past decade, there has been increasing concern over the hazards of fluorocarbons used as aerosol propellants, a topic reviewed by the author (D. M. A.)[7,13] and which will also be the subject of the next volume in the series. The composition of the aerosols discussed in the present volume excludes fluorocarbons and instead includes hydrocarbons, chlorinated solvents, and alcohols.

It is customary to use a combination of propellants and solvents in aerosol products, the former to generate pressure in the aerosol unit, and the latter to suspend and/or dissolve the ingredients dispensed by the unit. A retrospective examination of the preceding chapters in this volume, as well as the two previous volumes,[14,15] leads to the conclusion that the combination of hydrocarbons, methylene chloride, and ethanol has definite advantages over the use of fluorocarbons and other solvents such as trichloroethylene, isopropanol, and ketones. The reasons, based on the results of original investigations reported both in this volume and in the earlier volumes,[14,15] are as follows.

1. Hydrocarbons are less toxic than fluorocarbons – The concentration of trichloromonofluoromethane (FC 11) producing cardiac arrhythmia in the dog is 0.25%.[6] However, no arrhythmia is elicited with up to a maximal inspired concentration of 10% isobutane (Chapter 6), 10% butane (Chapter 5), or 20% propane (Chapter 4). This is the most important reason for the preference of isobutane over fluorocarbons; when aerosols are abused by inhalation, the cause of death is cardiac arrhythmia.[7]

A second difference between the two groups of propellants is related to their potential to depress cardiac function, specifically left ventricular dp/dt, stroke work, and cardiac output. Table 9.1 outlines the concentration of each propellant found to depress each of the three parameters. Among the fluorocarbons, FC 11 is the most toxic, causing depression of myocardial contractility at a concentration of 0.5%. Dichlorotetrafluoroethane (FC 114) ranks second, depressing contractility at a concentration of 2.5%, and dichlorodifluoromethane (FC 12) ranks third, the threshold concentration being 10%.

The order of the hydrocarbons, starting with the most toxic depressant to the heart, is as follows: 0.5% butane, 2.0% isobutane, and 3.3% propane. The hydrocarbon mixture A-46 is effective at a concentration of 2%, which is 4 times as high as the threshold for FC 11 (Chapters 4 to 7).

2. Halogenated hydrocarbons are less toxic than fluorinated hydrocarbons – Subacute inhalational exposure to methyl chloroform in mice indicates that this chlorinated solvent is unlikely to produce hepatotoxicity or pneumotoxicity (Chapter 1). However, similar exposure to fluorocarbons produces hyperplasia of the lymphoreticular system in the airways and interferes with bronchomotor function.[16]

3. Methylene chloride is less toxic than methyl chloroform and trichloroethylene – Among the three chlorinated solvents that have been tested so far on the canine heart, trichloroethylene is the most toxic, causing depression at the threshold concentration of 0.05%. The remaining solvents have higher threshold concentrations: methyl chloroform, 0.25%, and methylene chloride, 2.5%. In lower concentration (0.5%), methylene chloride even causes an improvement in contractility because of the release of catecholamines. There is a 50-fold difference in concentration between the most toxic and least toxic chlorinated solvents (Table 9.2). For the induction of cardiac arrhythmia, the two can be ranked as follows: 2.5% methyl chloroform, and 5.0% methylene chloride (Chapter 2). As a solvent, there is sufficient reason to prefer methylene chloride over the other chlorinated solvents. The potential for methylene chloride to form carbon monoxide[17] is not a serious deterrent, since brief exposure of human subjects to methylene chloride-containing aerosols did not produce an elevation of carboxyhemoglobin.[18]

4. Ethanol is less toxic than isopropanol – There is a definite advantage to ethanol over isopropanol, if it is necessary to use an alcohol as a solvent in aerosol products. The threshold concentrations which depress contractility are as follows: isopropanol, 1%, and ethanol, 7.5% (Table 9.2). The lower level of toxicity for ethanol is reflected by the observation that in lower concentrations (1.0%) there is even an increase in contractility (Chapter 3).

5. Ethanol interaction with methylene chloride and hydrocarbon propellants – The toxicological justification for using ethanol in an aerosol became apparent in the course of testing its interaction with other constituents (Chapter 8). The arrhythmogenic effect of methylene chloride was blocked by the concurrent administration of ethanol. Furthermore, the hypotensive effect of hydrocarbon mixture A-46 was reduced by ethanol.

6. Propellant-solvent interaction is less likely with hydrocarbons – The fluorocarbons, particularly FC 11, caused cardiac arrhythmia by sensitizing the heart to epinephrine.[19] Unrelated to this is the ability of ketones, specifically methyl isobutyl ketone, to potentiate the myocardial depressant action of FC 11.[15] The hydrocarbons, such as isobutane, do not show any interaction with ketones (Chapter 6).

7. Methylene chloride interaction with other solvents – The unexpected finding that the inhalational toxicity of methylene chloride in mice and dogs is reduced by the addition of toluene, methanol, and isopropanol raises questions relating to the mechanisms for interaction (Chapter 2). Only isopropanol has been examined in detail,[15] and an interaction with trichloroethylene has been excluded. Identification of the mechanisms for reducing the toxicity of methylene chloride is the topic for an anticipated forthcoming volume in this series.

TABLE 9.1

Summary of Cardiac Effects of Inhalation of Fluorocarbon and Hydrocarbon Propellants

Concentration % v/v	Effect on**	Hydrocarbons				Fluorocarbons		
		Propane	Butane	Isobutane	A-46 mixture	FC 11	FC 12	FC 114
0.50	a		−85	−21	−38	−219		
	b		−0.9*	−1.0	−0.8			
	c		−70*	−33	−38	−24		
1.00	a		−170	−19	−145*	−213*		0
	b		−1.2	−0.8	−1.7*			
	c		−80	−80	−130*	−52		−17
2.50	a	−84*	−280*	−583*	−175	−667		−222
	b	−0.5	−1.6	−2.6*	−2.1			
	c	−20	−100	−125*	−120	−129*		−92*
5.00	a	−229	−430	−583	−170	−906	−111	−334
	b	−1.6*	−2.1	−3.4	−3.4			
	c	−70	−120	−133	−150	−248	−24	−100
7.50	a			−812	−365			
	b			−5.8	−4.0			
	c			−263	−200			
10.00	a	−354	−635				−267*	
	b	−2.4	−3.2					
	c	−130*	−160				−104*	
20.00	a	−959					−456	−578
	b	−3.8						
	c	−230					−151	−183

*$P < 0.05$
**a = left ventricular dp/dt (mmHg/sec), b = stroke work (g·meter), and c = cardiac output (ml/min).

TABLE 9.2

Summary of Cardiac Effects of Inhalation of Chlorinated Solvents and Alcohols

Concentration % v/v	Effect** on	Chlorinated solvents			Alcohols	
		Methyl chloroform	Trichloroethylene	Methylene chloride	Isopropanol	Ethanol
0.01	a		+125			
	b		-0.2			
	c		0			
0.05	a	-72	-292*			
	b		-0.6			
	c	-17	-33			
0.10	a	-200	-323			
	b		-1.9			
	c	-55	-67*			
0.25	a	-393*	-542			
	b		-2.1*			
	c	-79*	-125			
0.50	a	-629	-724	+50	-104	
	b		-3.3	+0.3	-0.4	
	c	-129	-133	-10	-25	
1.00	a	-943	-1188	+75	-208	+94
	b		-5.7	-0.3	-0.9	-0.8
	c	-171	-275	-40	-50	-88
2.50	a			-300	-313	0
	b			-1.4	-1.5*	-1.1
	c			-160*	-84*	-100
5.00	a			-925	-417	-313
	b			-4.2*	-2.0	-2.1
	c			-560	-100	-150
7.50	a				-458	-406*
	b				-3.2	-2.9
	c				-150	-225
10.00	a					-656
	b					-3.6
	c					-275

*$P < 0.05$
** a – left ventricular dp/dt (mmHg/sec), b = stroke work (g·meter), and c = cardiac output (ml/min).

REFERENCES – Aerosol Formulations

1. **Hughett, P. D.,** Deodorizing aerosol sprays, German patent 1,645,144, 1970.
2. **Kaempfer, H.,** Propane-butane mixtures as aerosol propellants, *Aerosol Rep.,* 6, 178, 1967.
3. **Scott, R. J. and Werner, R. C.,** Propellant blends for aerosol deodorants, *Soap Chem. Spec.,* 42, 190, 1966.
4. **Mace, H. and Carrion, C.,** Technology of consistent aerosol foam dispensing without costly gadgetry, *J. Soc. Cosmet. Chem.,* 20, 511, 1969.
5. **Aviado, D. M.,** Toxicity of aerosol propellants in the respiratory and circulatory systems. X. Proposed classification, *Toxicology,* 3, 321, 1975.
6. **Aviado, D. M.,** Toxicity of aerosol propellants in the respiratory and circulatory systems. IX. Summary of the most toxic: trichlorofluoromethane (FC 11). *Toxicology,* 3, 311, 1975.
7. **Aviado, D. M.,** Toxicity of aerosols, *J. Clin. Pharmacol.,* 15, 86, 1975.
8. **Nakano, J. and Prancan, A. V.,** Effects of adrenergic blockade on cardiovascular responses to ethanol and acetaldehyde, *Arch. Int. Pharmacodyn. Ther.,* 196, 259, 1972.
9. **Schneider, F. H.,** Acetaldehyde-induced catecholamine secretion from the cow adrenal medulla, *J. Pharmacol. Exp. Ther.,* 177, 109, 1971.
10. **Watanabe, K. and Takesue, S.,** Effect of some hydrocarbon gases on egg-white lysozyme activities on different substrates, *Enzymologia,* 41, 99, 1971.
11. **Watanabe, K. and Takesue, S.,** Interaction of some gaseous hydrocarbons with egg-white lysozyme, *Hoppe-Seyler's Z. Physiol. Chem.,* 355, 184, 1974.
12. **Watanabe, K. and Takesue, S.,** Effect of some hydrocarbon gases on egg-white lysozyme activities on different substrates, *Enzymologia,* 41, 99, 1971.
13. **Aviado, D. M.,** Toxicity of propellants, in *Progress in Drug Research,* Jucker, E., Ed., Birkhauser Verlag, Basel, 1974, 365.
14. **Aviado, D. M., Zakhari, S., Simaan, J. A., and Ulsamer, A. G.,** *Methyl Chloroform and Trichloroethylene in the Environment,* CRC Press, Cleveland, 1976.
15. **Zakhari, S., Leibowitz, M., Levy, P., and Aviado, D. M.,** *Isopropanol and Ketones in the Environment,* CRC Press, Cleveland, 1976.
16. **Watanabe, T., Ratcliffe, H., Thompson, G., and Aviado, D. M.,** *Fluorinated Propellants for Aerosols,* CRC Press, Cleveland, in press, 1977.
17. **Stewart, R. D. and Hake, C. L.,** Paint-remover hazard, *JAMA,* 235, 398, 1976.
18. **Skory, L. K., Anthony, T., and Stevenson, M. F.,** Carboxyhemoglobin studies show methylene chloride safe in aerosol use, *Aerosol Age,* 20, 5, 56, 1975.
19. **Aviado, D. M. and Belej, M. A.,** Toxicity of aerosol propellants on the respiratory and circulatory systems. I. Cardiac arrhythmia in the mouse, *Toxicology,* 2, 31, 1974.

AUTHOR INDEX

I

J

K

L

M

N

O

P

R

S

T

U

V

W

Y

Z

SUBJECT INDEX

E

F

G

H

I

L

M

CRC PUBLICATIONS OF RELATED INTEREST

CRC HANDBOOKS

CRC HANDBOOK OF CHEMISTRY AND PHYSICS, 57th Edition

Edited by **Robert C. Weast, Ph.D.,** Consolidated Natural Gas Co., Inc.
This Handbook is the definitive reference for chemistry and physics and maintains the tradition that has earned it the reputation as the best scientific reference in the world.

CRC HANDBOOK OF CHROMATOGRAPHY

Edited by **Gunter Zweig,** U.S. Environmental Protection Agency and **Joseph Sherma, Ph.D.,** Lafayette College.
This two-volume set provides comprehensive information concerning chromatographic data, methods, and literature. It also contains a Compound Index that lists the more than 12,000 compounds referenced in this data collection.

CRC HANDBOOK OF LABORATORY ANIMAL SCIENCE

Edited by **Edward C. Melby, Jr., D.V.M.,** The Johns Hopkins University School of Medicine and **Norman H. Altman, V.M.D.,** Papanicolaou Cancer Research Institute, Miami.
This multi-volume reference is designed to provide a ready source of information on laboratory animals used in biomedical teaching, research, and testing. A cross-referenced cumulative index is included.

CRC HANDBOOK SERIES IN CLINICAL LABORATORY SCIENCE

Editor-in-Chief, **David Seligson, Sc.D., M.D.,** Yale University School of Medicine.
This unique Handbook Series offers a comprehensive reference source for selected areas of clinical laboratory science: Blood Banking, Clinical Chemistry, Hematology, Immunology, Microbiology, Nuclear Medicine, Pathology, Toxicology, and Virology and Rickettsiology.

METHODOLOGY FOR ANALYTICAL TOXICOLOGY

Edited by **Irving Sunshine, Ph.D.,** Cuyahoga County Coroner's Office (Ohio).
This volume, which replaces the Manual of Analytical Toxicology, is designed primarily for the detection of therapeutic concentrations of medications in biological fluids.

CRC "UNISCIENCE"™ PUBLICATIONS

BIOLOGICAL MONITORING FOR INDUSTRIAL CHEMICAL EXPOSURE CONTROL

By **A. L. Linch, M.S.,** Environmental Health, Everett, Pennsylvania.
This book reviews the direct quantitative analysis of expired air, body fluids, and tissue for the presence of hazardous substances and the effect of these substances on the target organs or tissues.

CADMIUM IN THE ENVIRONMENT, 2nd Edition

By **Lars T. Friberg, M.D., Magnus Piscator, M.D., Gunnar Nordberg, M.D.,** and **Tord Kjellstrom, M.E., M.B.,** the Karolinska Institute, Stockholm, Sweden.
This review updates the earlier review on cadmium carried out by the U.S. Environmental Protection Agency and the Department of Environmental Hygiene of the Karolinska Institute, Sweden.

CARCINOGENESIS TESTING OF CHEMICALS

Edited by **Leon Golberg, M.S., B.Chir., D.Sc., D.Phil., F.R.C. Path.,** Chemical Industry Institute of Toxicology, North Carolina.
This book reviews the evaluation of drugs for carcinogenic potential as an essential aspect of drug development and the various problems concerning science, ethics, and epidemiology that surround the evaluation.

THE CHEMISTRY OF PCB's

By **O. Hutzinger,** National Research Council, Halifax, Canada, et al.
This book presents a comprehensive summary of the chemistry of chlorobiphenyls.

CRC DRUG DEPENDENCE SERIES

Editor-in-Chief, **S. J. Mulé, Ph.D.,** New York State Narcotic Addition Control Commission.
The open-ended series on drug dependence was created to fill the obvious need for authoritative information within selected areas of drug dependence.

DRUGS AS TERATOGENS

By **James L. Schardein, B.A., M.S.,** Parke, Davis and Company.
Principles of teratogenesis, methods of laboratory assessment for determining the teratogenic potential of drugs, and their relevance to drug usage in the human are discussed.

EVALUATION OF AMBIENT AIR QUALITY BY PERSONNEL MONITORING

By **A. L. Linch, M.S.,** Environmental Health, Everett, Pennsylvania.
This book provides an overall view of the current air-sampling techniques available with critical comment based on unpublished discussion with specialists familiar with problems encountered in the field.

IMMUNOASSAYS FOR DRUGS SUBJECT TO ABUSE

Edited by **S. J. Mulé, Ph.D.,** New York State Narcotic Addiction Control Commission.
A select group of people actively engaged in the development or the evaluation of immunoassay techniques for the detection of drug abuse offer their findings in this comprehensive reference.

MERCURY IN THE ENVIRONMENT

By **Lars T. Friberg, M.D.,** The National Institute of Public Health, Stockholm, Sweden and **Jaroslav J. Vostal, M.D., Ph.D.,** University of Rochester School of Medicine and Dentistry.
This is an extensive review of the toxicology of mercury, dealing with the health aspects of contamination of the general and industrial environment with metallic mercury and different inorganic and organic mercury compounds.

METHEMOGLOBINEMIA: A COMPREHENSIVE TREATISE

By **Manfred Kiese,** Pharmacological Institute of Munich University, Germany.
The monograph deals comprehensively with all aspects of evaluated ferrihemoglobin content of blood.

CRC CRITICAL REVIEW™ JOURNALS

CRC CRITICAL REVIEWS™ IN ANALYTICAL CHEMISTRY

Edited by **Bruce H. Campbell, Ph.D.,** J. T. Baker Chemical Co.

CRC CRITICAL REVIEWS™ IN CLINICAL LABORATORY SCIENCE

Edited by **John Batsakis, M.D.,** University of Michigan Medical School and **John Savory, Ph.D.,** University of North Carolina.

CRC CRITICAL REVIEWS™ IN TOXICOLOGY

Edited by **Leon Golberg, M.B., B.Chir., D.Sc., D.Phil., F.R.C. Path.,** Chemical Industry Institute of Toxicology, North Carolina.

Please forward inquiries to CRC Press, Inc.